CAN YOU
SPEAK
VENUSIAN?

Books by Patrick Moore

A Handbook of Practical Amateur Astronomy
The Planets
A Survey of the Moon
Naked-Eye Astronomy
The New Look of the Universe
The Amateur Astronomer's Glossary
Amateur Astronomy
The Sun
Suns, Myths, and Men
How to Make and Use a Telescope, by Patrick Moore and
  H. P. Wilkins
Life on Mars, by Patrick Moore and Francis Jackson
Craters of the Moon, by Patrick Moore and Peter Cattermole
The Yearbook of Astronomy (annual)
Astronomical Telescopes and Observatories for Amateurs
Can You Speak Venusian

CAN YOU SPEAK VENUSIAN?

A GUIDE TO THE INDEPENDENT THINKERS

PATRICK MOORE

W · W · NORTON & COMPANY · INC ·
NEW YORK

Library of Congress Cataloging in Publication Data
Moore, Patrick.
    Can you speak Venusian?
    Bibliography: p.
    1. Astronomy—Curiosa and miscellany.
    I. Title.
B   QV52.M66 1973   001.9   73–8846
    ISBN 0–393–06394–1

Published simultaneously in Canada
by George J. McLeod Limited, Toronto

PRINTED IN THE UNITED STATES OF AMERICA

1 2 3 4 5 6 7 8 9 0

CONTENTS

ILLUSTRATIONS

FOREWORD

In 1969 I was invited to present a programme on British Broadcasting Corporation Television, "One Pair of Eyes", and I selected as my subject the theme of Independent Thought. The programme took many weeks to make and was, I think, a success—at any rate it has been re-broadcast several times.

It subsequently struck me that there was no book upon precisely this subject, and that much research work has gone unnoticed. Therefore I continued my investigations, and the present book is the result. It does, I feel, contain material which has not appeared before in collected form, and I hope that at least some of the ideas put forward may be of help to our glorious successors who will carry *Homo sapiens* toward the stars.

I have had much assistance. Many of those whose views are given here (notably Mr. Bradbury, Mr. Shuttlewood, the Rev. P. H. Francis, Mr. Pedrick, Mr. Norcott and Mrs. Parker) have read through the relevant sections and made most valuable comments and suggestions. The original BBC programme was produced by Simon Campbell-Jones, who has continued to be a tower of strengh. And, as always, I have had every help and encouragement from the publishers.

PATRICK MOORE

Selsey, May 1972

I

THE INDEPENDENT THINKER: LOOKING AT THE WORLD

"One evening many years ago in Glasgow, I noticed a series of round balls travelling round three walls of a room. Whatever was causing the images, about three or four inches in diameter? I looked and looked, and could discover nothing. Then I noticed that stars in the sky were fading, and that as they faded, my 'balls' disappeared! Stars were round! How is it that everyone thinks them pointed?"

So wrote Miss Margaret L. S. Missen, of Edinburgh, in a pamphlet published a few years ago at the very reasonable price of one (old) penny. It was by no means her only publication. Among others were *A New Theory of Gravity*, in which she pointed out that something much more powerful than conventional gravitation was required to prevent the Earth from tumbling through space; *The Moon*, in which she stated that according to astronomers the Moon's phases are due to the shadow of the Sun— clearly a ridiculous theory; and *The Sun Goes Round the*

Earth, the title of which speaks for itself.

Ideas of this sort do not appeal to professional scientists. I do not subscribe to them myself, though I speak as an amateur astronomer rather than as a professional. Neither can I believe that the world is flat, that our destinies are ruled by astrology, that we are being constantly visited by people from afar who whiz down in their flying saucers, or that the whole concept of Darwinian evolution is a foul plot intended to mislead the up-and-coming generation. I have no hopes of inventing a perpetual motion machine, or of preventing earthquakes by means of splitting the Moon in half and sending one hemisphere out to the other side of its orbit. I do not share the fear that the world's axis will suddenly tilt, thereby causing a universal flood compared with which Noah's would seem like the trickle from a bath-tap. But this is, no doubt, because I am a cosy convention-alist—and my misgivings do not in the least lessen my admiration for the Independent Thinkers who are the subjects of my present book.

Note that I say 'Independent Thinker', and not 'crank'. The difference is important. The Independent Thinker is a genuine, well-meaning person, who is not hidebound by convention, and who is always ready to strike out on a line of his own—frequently, though not always, in the face of all the evidence. He is ready to face ridicule; he believes himself to be in the right, and he cannot be deterred. In some respects he is a rather special kind of person, though generally speaking he is conventional enough except in his one particular line of thought. He may or may not be scientifically qualified. In the following pages you will meet some people who have no claims to academic emin-ence; but you will also meet others who have. All share the wish to inquire, and—this is the vital fact—all are anxious to do something really useful.

There are, of course, unconventional people who are decidedly different. I propose to do no more than mention

10

them here, because I do not like them and I wish to have nothing whatsoever to do with them. Some religious (or pseudo-religious) cults come under this heading, and I think it is clear that they can do immense harm by influencing gullible people and even breaking up families. There have been several well-publicized cases of this in recent years; I have known some myself, and the circumstances were most unpleasant. I believe that people in Britain, for example, are too tolerant of these cults. Unfortunately there are difficulties in the way of controlling them, because in the ordinary way they are careful to keep on the right side of the law; but they are not at all nice— and it is people of this sort whom I would label either as cranks or as crooks (sometimes both). The less I see of them, the better I am pleased.

So let us concentrate upon the true Independent Thinker, and see just how his mind works.

The first thing to remember is that our knowledge of the universe is really very limited. Most of us agree that we live on a planet almost 8,000 miles in diameter, moving round the Sun at a distance of 93 million miles; that the Sun is a star, and that in our Galaxy we find something like 100,000 million stars. Beyond lie other galaxies, of the same order of magnitude. With modern instruments we can analyze starlight, and find out what substances are there; we can send men to the Moon and dispatch rocket probes to our neighbour planets. But when we ask just how the Earth and the rest of the universe came into being, we have to admit that we are totally ignorant. The atoms and molecules that make up you, me, this book, the kitchen sink, Aunt Emily, the Sun and the galaxies must have come from somewhere: but where? If we begin with a universe filled with homogeneous gas, we can work out an evolutionary sequence; but we cannot decide how the gas got there in the first place. Neither can we understand the nature of time, and we are very uncertain about the origin

11

of life itself. In fact, modern science is strong on details, weak on the essential fundamentals. Moreover, one cannot prove a negative. What the Independent Thinker usually does is to go to the grass roots and question *everything*. This is why he cannot be refuted by conventional argument.

Can you prove that the Sun will rise in the east tomorrow morning? I certainly cannot. I think it will, because all the evidence points that way; but not until the Sun pokes its head over the horizon at dawn will I be able to give you a final answer. Rigel, the brilliant white star in the constellation of Orion, is believed to be 900 light-years away; that is to say, its light has taken 900 years to reach us. Therefore we are seeing it not as it is today, but as it used to be 900 years ago, at about the time when King Harold was on the receiving end of a Norman arrow. Nobody can prove that Rigel exists now. All we can say is that it did exist when the Battle of Hastings was fought. Our knowledge of the universe—excluding our immediate locality—is bound to be very much out of date.

But, of course, if the Earth is a flat disk, if the Moon is a small body covered with plasticine phosphorus, if the Sun is cold...where do we go? If all our geological methods of dating are absurdly wrong, Man could be a newcomer to the terrestrial scene, and *Homo sapiens* might have popped up out of a metaphorical trap, with none of this arbitrary nonsense about evolution from Eocene primates. It is no use questioning a part of the evidence. One must reject all of it, with no holds barred. This is what the Independent Thinker does; and he is ready to accept the consequences.

Bear in mind that the Independent Thinkers of the past have been given similar short shrift. Copernicus, in the mid-sixteenth century, published a book in which he claimed that the Earth moves round the Sun instead of vice versa. He was immediately attacked, not only by religious leaders such as Luther (which was predictable) but also by scientists. "Who is this fool who wants to turn the world upside-down?" was one comment. The irony is that most of Copernicus' theories were wrong. True, he took the important step of removing the Earth from its proud position in the centre of the universe and putting the Sun there instead; but in all other ways he accepted the ancient ideas—notably the obsession with perfectly circular paths. Yet it was his one correct idea which drew the wrath of his contemporaries down upon his head. (In fact, he avoided trouble by the simple expedient of postponing publication until he was dying. Some of his followers were less prudent, and at least one of them was burned at the stake—a fate which, I am glad to say, is not likely to overtake any modern critic of science.)

Darwin, too, had his battles to fight; they took some time to win, and according to the Evolution Protest Movement they are not won yet. When the great chemist Dalton published his periodic table of the elements, someone made the caustic suggestion that he would find it more profitable to arrange the elements alphabetically . . . and so on.

In more modern times, we have men who are unquestionably of the first rank in their fields of research, and yet have proposed ideas which do not meet with general acceptance. Do you remember John Logie Baird? He was undoubtedly the pioneer of television; even though his system was not ultimately adopted as he had originally planned, he had a proper TV set working earlier than anybody else. But in his younger days he also had the brilliant idea of making diamonds, by putting a rod of carbon into a cylinder and heating it violently by means of an electric current. The predictable result was a violent explosion which not only shattered his apparatus, but also fused all the lights in the building—and his employers were by no means enthusiastic. Dr. Iosif Shklovskii, one of Russia's leading astronomers and a man who has done work of

13

fundamental importance in astrophysics, believes that the two tiny satellites of the planet Mars are artificial space-stations, launched by the Martians for reasons of their own. And Dr. Ernst J. Öpik of Armagh, Northern Ireland, another leading astronomer, has correlated sunspot activity with political revolutions. In his book *The Oscillating Universe*, he says that this link "cannot possibly be due to coincidence".

(There was another eminent astronomer of our own century, T.J.J. See, who firmly believed himself to be immortal. When he passed on to better and higher spheres, which he duly did at an advanced age, he must have had a tremendous shock.)

Under exceptional circumstances, unconventional ideas can have the benefit of official backing. I have some inner knowledge of this, because at one stage during the war I was associated with a department in Britain which was generally nicknamed the Department of Bright Ideas. We had some remarkable suggestions put to us, of which my favourite was a foolproof plan for defeating the German radar; all one had to do was to build a raft the size of the British Isles and then float it in the North Sea, causing total confusion to the Luftwaffe. I lost count of the number of perpetual motion machines submitted to us (we finally began asking for working models, after which the supply of inventors dried up). But there was also Project Habbakuk, which involved chipping an iceberg off the North Polar cap, towing it down to the Channel, and using it as an aircraft-carrier, filling up bomb-holes with water which would promptly freeze and retain a smooth surface suitable for take-off and landing. This project had, for a time, the full backing of the War Cabinet!

In my personal investigations into the world of Independent Thought, I have been impressed with the charm, the courtesy and the patience of those concerned. They are so totally unlike the religious cranks, the money-grubbers, and the more extreme child psychiatrists whose doctrines have caused so much havoc during the past couple of decades.

I must repeat what I said earlier: I am a conventionalist, and also somewhat sceptical, so that in a way I am acting as devil's advocate. Now and then the Independent Thinker may be right when others are wrong. More often he is grasping hold of a totally false idea. But let him have his say; and I invite you to join me in the fascinating realm of the flat earth, the cold sun, the fallacy of evolution, and sundry visitors from Venus. I hope that you will enjoy the journey.

2

BETTER AND FLATTER EARTHS

As I write these words, I am sitting in my quiet study in Sussex, England, looking out over the rose-garden toward the belt of trees which shields us from the sea. There is the gentle breeze so familiar in Selsey, but nothing more. Yet it has been claimed that if the Earth were spinning round, as conventional scientists claim, there would be a howling gale all the time.

To see just how this theory works, we must go back almost two thousand years—in fact to the second century A.D., when the most famous scientist in the world was Claudius Ptolemæus, better known as Ptolemy. We know very little about his life, except that he flourished from around A.D. 120 to 180; that he lived in Alexandria, and that he belonged to the Greek school of thought. He was an expert astronomer and mathematician, and also a geographer; his map of the known world was remarkably good, even though he did join Scotland on to England in a sort of back-to-front position. Also, he wrote books. Much of our knowledge of ancient science is due to him, because, by a miracle, his books have come down to us—even though only by way of their Arab translations.

(Ironically, it seems that the great Alexandrian Library, which contained priceless books dating from the very early days, was destroyed during the time of Arab supremacy. There is a legend that they were deliberately burned by order of the Caliph—because if they contradicted the Koran they were heretical, while if they agreed with it they were superfluous. The story is decidedly dubious, and it may well be that the Library was gradually dissipated by neglect. However, the end result was the same: all the books were lost.)

A few earlier Greeks, such as Aristarchus of Samos, had taught that the Earth is a planet moving round the Sun, and that it rotates on its axis. Ptolemy could not bring himself to accept this secondary rôle for the Earth, even though he was quite prepared to believe that the world is a globe. His reason was quite straightforward. If the Earth is whirling round, and the atmosphere is not whirling with it, the result will be constant, violent wind—just as you can experience today if you stand up in an open car which is travelling along the motorway at 60 m.p.h.

For many years I was naïve enough to believe that this idea had died a natural death, and it was with some surprise that in 1957 I read some words in a book called *Looking at the Stars*, written by a professional astronomer —Dr. Michael W. Ovenden, Fellow of the Royal Astronomical Society. Dr. Ovenden was discussing the craters of the Moon, and pointed out that no craters of the same kind are likely on our own world, because "any large lunar-type craters on Earth would in a few million years be rubbed away by friction with the atmosphere as the Earth rotates underneath it". It is an intriguing idea. Somehow, however, the murmuring breeze now passing through my rose-garden leads me to believe that both Ptolemy and Dr. Ovenden were wrong.

17

The moving Earth has also come under fire from other researchers. Again I quote Miss Missen—without apology, because her views are so striking. In *The Sun Goes Round the Earth*, she wrote:

"If the Earth did move at a tremendous speed, how could we keep a grip on it with our feet? We could walk only very, very slowly; and should find it slipping rapidly under our footsteps. Then, which way is it turning? If we walked in the direction of its tremendous speed, it would push us on terribly rapidly. But if we tried to walk against its re-volving ——? Either way we should be terribly giddy, and our digestive processes impossible."

Equally forthright was Mme. Gabrielle Henriet, whose book *Heaven and Earth*, published in 1957, is a master-piece of Independent Thought. I shall return to some of her theories later—notably her revelation that the sky is

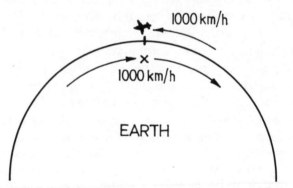

Fig 1 The non-rotation of the Earth according to Mme. Henriet

solid—but for the moment it will be enough to give her disproof of the rotation of the world. She begins by point-ing out that the rate of spin given by astronomers is 1,000 kilometres per hour. Modern aircraft can attain this speed: but "an aircraft flying at this rate in the same direction as that of the rotation could not cover any ground at all. It would remain suspended in mid-air over the spot from which it took off, since both speeds are equal. There would, in addition, be no need to fly from one place to another situated on the same latitude. The aircraft could just rise and wait for the desired country to arrive in the ordinary course of the rotation, and then land; although it is difficult to see how any plane could manage to touch ground at all on an airfield which is slipping away at the rate of 1,000 kilometres per hour. It might certainly be useful to know what people who fly think of the rotation of the Earth."

Speaking as an ex-Bomber Command flyer—1940-45, practically in the Stone Age of aviation—I can only admit that I am speechless; but even if I cannot agree with Mme. Henriet, I have tremendous admiration for her ingenuity. In a British television programme called 'One Pair of Eyes', I was very anxious for her to join me; and she declined only on the grounds that if she faced a TV camera, her false teeth would fall out. I would be the last to deny this possibility (in fact, nothing appeared to be more probable), but I was very sorry about it.

Mr. John Bradbury also believes in the non-rotation of the Earth, and has put forward his theories on the radio, on television and in lectures to Universities. However, his view of the universe as a whole is so remarkable, and so interesting, that it deserves a separate chapter, and I propose to defer discussion of it for the moment.

Before passing on, I must pause to give another way of proving that the Earth is motionless. It was outlined at a meeting of the Flat Earth Society some years ago, and it is pleasingly direct. Go out at night-time, point your camera at the stars, and make a time-exposure for, say, a quarter of an hour. When you develop the plate or film, you will see numerous star trails. These trails will be hard, sharp lines. But if the Earth were moving, the trails would be blurred. Just you try taking a time-exposure out of the window of a moving railway-carriage!

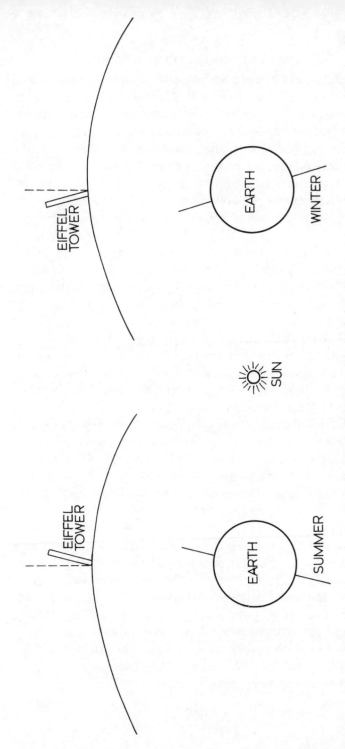

Fig 2 Mme. Henriet's theory of the seasons

When this idea was explained, I did tentatively suggest that the trails might be due to the actual rotation of the Earth and not to the individual movements of the stars. Naturally, this objection was brushed aside with the contempt that it deserved. And to return to Mme. Henriet, it is maintained in her book that the changing seasons cannot be due to the tilt of the Earth's axis, as astronomers say. If the axis pointed one way in summer and the opposite way in winter (see diagram), then very tall buildings, such as the Eiffel Tower, would sway drunkenly from side to side . . .

All the Independent Thinkers so far mentioned are sober researchers, concerned only with pure science. When we come on to the flat earth theory as a whole, it is true that we do tend to touch upon the realm of religion, and some flat-earthers are also Biblical Fundamentalists. However, speaking as an aspiring scientist, I do not propose to discuss the religious aspect here. I must gloss over the comment that it would be impossible for four angels to stand at the corners of the Earth, as stated in the Bible, unless the world were square or at least rectangular.

In fact, the ancient Egyptians did believe in a universe which took the form of a rectangular box, with the longer sides running north-south, and with a flat ceiling, supported by pillars at the cardinal points. The pillars were joined by a chain of mountains, and below the crests of the peaks lay a ledge containing the celestial river Ur-nes. The boats carrying the Sun and other gods sailed along this river. When a boat came to a corner, it described a graceful right-angle and continued blithely on its way. Combined with these strictly scientific ideas were various religious ones; in some parts of the Nile Delta it was thought that the heavens were formed by the body of a goddess whose name was, appropriately, Nut, and who was suspended permanently in what must have been an uncomfortable as well as an inelegant position. Egypt lay in the centre of

21

the flat Earth, and was surrounded on all sides by a boundless ocean.

It is tempting to dwell upon these old theories. I also like that of Vedic lore, in which the centre of the Earth was marked by a tall mountain, around which moved the celestial bodies in horizontal paths at different heights; the sky was, of course, solid (shades of Mme. Henriet). Other Indian thinkers believed the Earth to be carried on the shoulders of elephants, which were in their turn supported on the shell of a huge turtle swimming in the sea. I would feel rather sorry for the turtle, and it is hard to avoid the conclusion that it would end up by being turned into some kind of turtle soup; but it is time to come back to the present century, and to consider one of the most interesting and long-established societies of Independent Thought —the International Flat Earth Society.

It has been in existence for a long time, with its headquarters in Britain; but a few decades ago the kernel of flat-earth belief was Zion, Illinois, where Wilbur Glenn Voliva ruled his community with an iron hand. He believed the world to be shaped like a pancake, with the North Pole in the middle and a wall of ice all round. There is no South Pole, but fortunately the icy barrier prevents ships from sailing over the edge and tumbling into Hades—below which, incidentally, is a bargain basement area inhabited by the spirits of a race of men who used to live on Earth before the arrival of Adam and Eve.

I never met Wilbur, who died in 1942; but I did know Samuel Shenton, who may be described as the Isaac Newton of Flatearthology. He joined me in the British television programme 'One Pair of Eyes', and his death, early in 1971, was a sad moment. By profession he was a sign-writer; but his theories made him world-famous, and he was even referred to in a broadcast made by Colonel Frank Borman from the Apollo space-craft during the lunar flight of Christmas 1968. He was utterly sincere and completely dedicated. Developments such as artificial satellites and journeys to the moon caused him no more than a few moments' misgivings before he was himself again. His great regret was that he had so few followers. As the organizer of I.F.E.S., he did not spare himself—and he continued his crusade in spite of indifferent health; he was by no means an old man at the time of his death. From his home in Dover he continued to write, to lecture, and to make occasional appearances on television.

Before going any further, let me give the official I.F.E.S. answers to just a few of the questions which spring automatically to the mind. First, just how can one prove, by everyday observation, that the Earth is a globe? Let us try.

Questioner. How do you explain the fact that if you watch a ship sailing over the horizon, you will see the hull vanish first, followed by the funnels and finally the smoke (if any)?
Flat Earther. Have you ever actually *seen* this happen?

(Collapse of questioner, who, in 99 cases out of a hundred, hasn't. But let us assume that he is the hundredth case.)
Questioner. Well, yes, I have, actually. How do you explain it?
Flat Earther. By the refraction of light. If you watch the phenomenon several times, sooner or later you will see the entire ship apparently suspended in the air above the horizon—and I imagine you don't believe in anti-gravity?

(This is a completely irrefutable argument, and the questioner has no choice but to start bowling on an entirely different wicket.)
Questioner. Aircraft can circumnavigate the Earth. Fly east—or west—and eventually you will come back to your starting point. This couldn't happen with a flat world.
Flat Earther. Of course it could. The Earth is shaped like a gramophone record, and all you have done is to complete a circuit round the central North Pole.

Questioner. Well, then, try going due south. On your theory, you would go over the edge.

Flat Earther. But no. One cannot go over the edge, because there is a wall of ice in the way.

Questioner. The compass direction—

Flat Earther. You're falling into the elementary trap of supposing that a compass needle always points north. This is not true. Near the periphery of the biscuit-shaped Earth, all compass directions are distorted, and this is why some explorers have deluded themselves into believing in a South Pole.

Questioner (after a baffled pause). Look, men have been in space, and have seen the Earth as a globe. They have even produced photographs.

Flat Earther. Nobody has ever seen the Earth as a sphere; all that the space-men have been able to do is to see wider areas of the world at any one time, which is quite understandable. I believe you also consider the Moon to be a globe—but if you care to look at it this evening, you will see that it appears as a flat disk.

Questioner. That doesn't explain the photographs.

Flat Earther. Fakes, produced by reactionary scientists in order to conceal the truth about the shape of the Earth.

At this point the questioner usually gives up and suggests adjourning to the nearest hostelry, where he will restore his shattered morale by drinking several stiff whiskies. There are, in fact, only two points upon which the Flat Earther cannot sound convincing. One is his inevitably evasive reply to questions about what the underside of the Earth is like, and what lies below *that*. The other is his contention that orthodox authorities are indulging in a campaign of suppression. Let me add that not all Flat Earthers think this; but one has to admit that even some of the most charming and patient Independent Thinkers have inner feelings that they are being "got at".

THE FLAT EARTH SOCIETY

President :
W. MILLS,
7 Vale Grove,
Finsbury Park, N.4.

Organizing Secretary :
S. SHENTON,
22 London Road,
Dover.

o o o o o o

The International Flat Earth Society has been established to prove *by sound reasoning and factual evidence* that the present accepted theory, that the Earth is a globe spinning on its axis every 24 hours and at the same time describing an orbit round the Sun at a speed of 66,000 m.p.h., is contrary to all experience and to sound common-sense.

In ancient times the Earth was regarded as plane, and this is expressed in all literature up to a few hundreds of years ago. The theory has fallen into disfavour, owing mainly to the dogmatism of modern science and popular education in schools, which leads to prejudice in favour of the globular theory from the start.

It is always a pity to allow false theories to pass unchallenged, and it is hoped that the Flat Earth Society will do much to undo the harm that has been caused. Remember that the truth of the plane figure of the Earth can be shown *by irrefutable evidence*, and anyone who is interested in becoming a member is asked to contact the President or the Organising Secretaryl In future, it is hoped to hold regular meetings of the Society.

DECEMBER 20TH, 1956.

Fig 3 Circular issued by the Flat Earth Society

Following a period of comparative inactivity, the British branch of the I.F.E.S. was revived in 1956, with a good deal of publicity. I reproduce here its official leaflet. ("Secretary1", on the penultimate line, is presumably a misprint; the 1 should be a full-stop.) On November 23 a meeting held in Finsbury Park at the home of the President, Mr. W. Mills, put the Society back on a really firm footing. Various abstruse papers were given; one, by Mr. Shenton himself, dealt with aerodynamical problems, and was greeted with great respect. Mr. Shenton pointed out that if gravity ended at a height of nine miles, as some scientists had maintained, then a parachutist coming down from a great height would miss the Earth altogether. Where he would go then remained something of a mystery.

I left the meeting in a mood of deep thought. Less than a year later—to be precise, on October 4, 1957—the Space Age started, not with a whimper, but with a very decided bang. Sputnik I sored aloft from its launching base in the Soviet Union, and sped round the Earth, sending back its famous "Bleep! bleep!" signals, and decisively putting paid to a suggestion made not long before that the whole concept of space-travel was utter bilge. This is not the place to discuss the wider implications of the Russian satellite. But what of its effect upon the Flat Earthers?

For a few brief days they were disconcerted—but, fortunately, not for long. A calm statement from the Society pointed out that the mere fact of a satellite moving above us did not show the world to be round. It could equally well be the more plausible flat disk, and the satellite would move in much the same way as the Moon, though at a lower level. This satisfied the dedicated enthusiasts. On the other hand, it does seem that the advent of the satellites was probably the main factor in preventing the Society from mushrooming as its organizers had hoped it would do. It is very greatly to their credit that they did not give up. They continued writing, lecturing, and arguing persuasively. They convinced a few people, but as the years went by, and the first tiny satellites were superseded by massive rocket probes to the Moon and beyond, even Mr. Shenton had to admit that the wicket had become very sticky indeed.

This was the situation in late 1968, when Apollo 8, carrying Astronauts Borman, Lovell and Anders, departed on the first voyage round the Moon. It is a tribute to the esteem in which Mr. Shenton was held that Colonel Borman actually referred to him in a broadcast made from the space-ship while it was between the Earth and the Moon. Not, of course, that Colonel Borman believed the world to be flat; it would have been rather difficult for him to do so, particularly at that moment!

A few months later Mr. Shenton joined me for a television discussion. It was, in fact, the last time I met him; and the views he expressed then must be regarded as his final word on the subject, because he did very little further lecturing or writing (mainly because of ill-health). So let us summarize what he laid down, eloquently and with total sincerity.

Originally the Earth was heaved up out of the waters. The North Pole marks the central point; the 60,000-mile periphery, coated with a vast barrier of ice, is what we conventionally call the South Pole. It is this barrier which confines us to the Earth and stops us from being in any danger of falling off. What lies on the underneath of the flat Earth remains mysterious, but there is no reason to doubt that it is highly complicated; it is impossible for us to compute the full extent of the Earth, because we can examine only that part of it which is surrounded by the wall of ice. The Earth may even be infinite, stretching out indefinitely. The Moon is a very small body, moving in an east to west direction; each night it sets later by 28 minutes or thereabouts, so that it shows us different aspects, and produces what astronomers call the phases, from new

27

to full. The Sun is larger, with a diameter of about 32 miles. The Sun's distance is less than 3,000 miles, as is shown by experiments carried out in South America. In parts of this continent, the true latitude lines show that there is a point at which latitude 45 degrees crosses the equator. (I admit to finding this argument a little hard to follow, but all I can do is to explain what Mr. Shenton explained to me.) From here you can get a triangulation to the conventional equator, 3,000 miles away. If the Sun is overhead at this moment, then the optical distance must be equal to the distance of the baseline upwards: that is to say, 3,000 miles.

As for the astronauts ... well, they simply went out in an egg-shaped orbit, and the photographs they brought

Fig 4 Apollo paths, according to Mr. Shenton

back were distorted due to the angle from which they were taken. The remaining photographs were subsequently faked by unscrupulous propagandists to disprove the true theory that the Earth is a flat plane.

And finally: the entire universe consists of the Earth. Certainly we lie on a great water base; the water seeps right through the Earth's disk, producing our present oceans and springs. With regard to greater distances, it is hard to speculate. There may well be a series or a stream of "heavens", made up of enclosed spaces and perhaps even inhabited.

What can one say? Mr. Shenton's ideas were not the same as those of the Flat Earthers of the early twentieth century—headed by Lady Blount, wife of Sir Walter de Sodrington Blount, who described herself as a mathematician, astronomer and lecturer as well as an explorer, and who believed the world to be flat and of strictly limited extent. However, both treated the conventional globe theory with contempt; both built up organizations to spread their theories; and it may be that in each case the empire was dependent upon its leader. I asked Mr. Shenton about the future of the Flat Earth Society, and he was decidedly pessimistic. "It will die," he said, "just as at the turn of the century the great organization set up by Lady Blount died. This is deliberate suppression on the part of orthodox scientists."

And yet—I wonder!

At the moment it would be idle to deny that the International Flat Earth Society is going through one of its leanest periods. It has very few members, no organization, and no leader to match Lady Blount, Wilbur Glenn Voliva, or Samuel Shenton. But somehow I have the feeling that it may survive, just as astrology has done; and frankly I hope it does. Those who believe the world to be shaped like a pancake are among the most attractive of the really Independent Thinkers.

29

3

HOLLOW EARTHS AND
SOLID SKIES

In the year 1823 a remarkable proposition was put to the Congress of the United States of America. It was due to Captain John Cleves Symmes, who had had a distinguished career as a soldier but who had given up his military duties in the noble cause of science. He was convinced that the Earth was made up of five concentric spheres, in each of which people lived. The way to the interior was through a wide opening at the North Pole, and there was a corresponding tunnel in the far south. Symmes wanted official backing for an expedition to the North Pole to see just where the hole began.

Congress listened politely, but with some scepticism. When the vote was taken, twenty-five members voted in favour of the Symmes expedition. It was encouraging, but not enough; and Symmes died six years later, the North Pole still unreached.

Certainly this was a fascinating theory, and at the time it was difficult to check. No scientific arguments would suffice, for the excellent reason that Symmes—like so many Independent Thinkers of later years—questioned *everything*, and there was no foundation upon which his critics could build. All one can really say is that his petition was no wilder than others which have been put before Congress and, for that matter, before the hallowed House of Commons—and have been passed.

There the matter rested until 1913, when another American, M. B. Gardner, came out with a rather similar theory. It was no real advance on Symmes', and I mention it here only because Gardner did have the extra notion that the wonderful coloured displays of lights known as auroræ are due to the radiance from an inner sun which shines out through the polar openings. The notion of an inner sun was revolutionary, but again the keystone was "the hole at the Pole", and when Gardner's resolution was put to the test he failed lamentably. Only a few years after he launched his theory on to the waiting world, there came the announcement that Admiral Richard Byrd had flown right across the North Pole, without seeing a tunnel of any kind. Latterly there have been doubts about Byrd's route, but in any case Gardner was discouraged. Apparently he conceded defeat, and nothing more was heard of him, though I believe he lived on until shortly before the outbreak of Hitler's war.

So far as I know, there are no modern followers of Symmes or Gardner. On the other hand, there is still a society dedicated to the idea that the Earth is hollow, and that we live on the inside of it. From Britain, Australia is somewhere overhead, and the Sun is in the middle of the Earth, surrounded by a crystal sphere upon which the stars are etched. This idea is particularly notable inasmuch as it led to a significant rocket experiment not so very long ago —in fact, in 1933. But before describing the so-called Magdeburg Experiment, I must pause to say something about Cyrus Teed, because he was really the originator of the whole scheme.

Teed was born in New York State in 1839, but it was only in 1869 that he had his great revelation. It came in the form of a vision, and was to the effect that the entire universe is made up of the hundred-mile-thick shell of the Earth. We live in the interior. Beyond, there is absolutely nothing. We cannot see Australia above us because of the density of the atmosphere, so that there is no hope of taking a telescope, pointing it upward, and seeing kangaroos prancing about. We cannot even see the real Sun, which is half bright and half dark. We have instead a sort of ersatz Sun, which is the reflection of the central one; when the real Sun has its dark half turned towards us, we have our period of night. There is no need to bother about the Moon, which is merely a reflection of the Earth.

Cyrus took some time to prepare his onslaught. Then, in 1870, he published his magnum opus, using the pseudonym of Koreshan, which is the Hebrew for Cyrus. His theory became known as Koreshanity Cosmology—and it still is.

Undoubtedly he expected to be hailed as the greatest genius of all time. Alas, his book hit the world with all the impact of a feather falling on to a piece of damp blotting-paper. Cyrus (apologies: Koreshan) was bitterly offended, and said so in no uncertain terms. Indeed, he went on saying so right up to the time of his death in 1908; and the magazine supporting his theory—appropriately entitled *The Flaming Sword*—went on appearing until after the end of the last war. The final issue known to me came out in 1948, though the periodical may have continued for slightly longer. But in the meantime, Space Research had stepped in—not to the detriment of Koreshanity Cosmology, surprisingly enough, but to provide and finance a test!

Again I must digress slightly; I hope that you will bear with me.

The idea of space exploration is very old indeed, and

there was even a book about a lunar journey written by a Greek satirist, Lucian, as long ago as the second century A.D. (In this book, the heroes of the story are swept to the Moon, together with their ship, by a violent waterspout. When they arrive, they are in time to take part in a war between the King of the Moon and the King of the Sun with regard to who shall have first rights on Venus—the planet, I mean, not the goddess.) The first idea which was not outrageous was Jules Verne's, produced in 1865, and involving a space-gun; in the famous novel *De la Terre à la Lune* the travellers were put into a hollow projectile and fired off to the Moon at a speed of seven miles per second. (If you have never read this book, I recommend you to do so. It is still quite fascinating, even though so dated.) But it was the Russian schoolmaster Konstantin Eduardovich Tsiolkovskii who first proposed using rocket propulsion; he issued his pioneer papers at the turn of the century, at about the time when Orville and Wilbur Wright were making their first hops in their primitive *Flyer*. Later, in the 1920s, the first liquid-fuel rockets were sent up. Priority must go to Robert Hutchings Goddard, in America; but much greater publicity surrounded the German group based near Berlin. There, a team including no less a person than Wernher von Braun established an experimental launching ground, and set up what became known as the Raketenflugplatz or Rocket Flying Field.

Initially they had a few successes, and a great many failures. Any rocket which went up for a few tens of feet without fizzing, exploding or diving back to terra firma was reckoned a triumph. Unfortunately money was short, and rockets, even squibs of the Raketenflugplatz type, were not cheap. The experimenters believed themselves to be on the right track—and indeed they were; their work led first to the V.2s of Peenemünde, then to the high-altitude scientific rockets, and finally, in 1969, to the Moon. But lunar flight seemed a long way off in the years before the war. What the German group needed was money—as much of it as they could raise.

Subsequently, of course, the Nazi Government stepped in and took over the entire project, transferring it to Peenemünde. But this was not so in 1933, just before the late unlamented Herr Hitler came to power. And this is where we come back to Koreshanity Cosmology (you were probably wondering when I would get round to it).

Magdeburg, with a reputation for progressive thinking, had of course a City Council. One member of it was named Mengering. He was an engineer; nobody seems to know much about him, and I imagine that he has long since passed on to better and higher spheres, but evidently he was an Independent Thinker with a strong will and a considerable amount of persuasion. He had read Teed, and had accepted all he said. But how could the theory be proved? If it could be substantiated, Magdeburg would become the leading scientific centre of the world—and, of course, Herr Mengering would achieve lasting fame.

Therefore, he cudgelled his brains to see what could be done, and he turned to the inventors at the Raketenflugplatz. Here, surely, was the solution. Teed had been unable to carry out practical experiments; but now, in the enlightened Thirties, it should be possible to launch a rocket powerful enough to give a decision one way or the other. Very well, then. Dispatch your rocket vertically upward. If it crosses the inside of the hollow Earth and arrives among the sage-bushes of Australia or the hot springs of New Zealand, then Koreshanity Cosmology will have been vindicated.

Mengering went into action, and called up Rudolf Nebel, one of the leaders of the German Society for Space Research. He came to the right man. Nebel's subsequent rôle in interplanetary affairs was not very creditable; he finally threw in his lot with the Nazis, and in various other ways also he showed that he was emphatically not to be trusted. However, he was adept at grasping opportunities, and he was suitably enthusiastic. If the City Council would provide the money, the Society would build and launch the rocket.

I have talked to people who were personally concerned in the whole affair, and who even saw the eventual launchings. Frankly, I have yet to find anyone who had the slightest faith in a hollow Earth; but a sum of several thousands of marks was not to be despised, and arrangements were made. The money was handed over, and construction began. I imagine that this is the only time that official funds have been made available to test a theory which is so far beyond the bounds of conventional science.

The plans were ambitious, and involved a rocket which would be more powerful than anything built before. Unfortunately the City Council was decidedly impatient, and even though the funds provided were essential to the project, there was still not enough ready cash. Refusing to be daunted, the pioneers went ahead. By March the first tests began, and resulted in a most unsatisfying explosion. Further major explosions accompanied the subsequent tests, but still the experimenters pressed on, and on June 9 came the first real firing. Alas, the rocket did not depart for the Antipodes. To be precise, it got no further than the top of its launching rack, after which it sank gently back to its starting position and refused to budge. A second attempt on June 13 was slightly more successful, since on this occasion the peak altitude was no less than six feet. The supreme moment came on June 29, when the rocket really did clear its launching paraphernalia. The trouble was that instead of being vertical, the trajectory was horizontal, and the journey ended in a belly-flop at a distance of roughly a thousand feet.

That was the end of the Magdeburg Experiment. It was also practically the end of the Society. Hitler became Führer; the Raketenflugplatz was closed, and before long most of the leading workers had been "taken over". The road to Peenemünde and the V.2 rocket weapon was opening. But since this is unconnected with the Independent Thinkers, it has no place here; and the entire episode remains as a monument to unorthodoxy—though, let it be repeated, neither von Braun, Nebel nor the other rocket men were disciples of Koresh.

For many years I thought that the hollow globe theory had died, but I then found that I was wrong. At Garmisch-Partenkirchen, in Germany, there arose the Society for Geocosmic Research, headed by Helmut K. Schmidt and P. A. Müller-Murnau. Their ideas were not identical with Koresh's, but were along the same lines. The main difference was that instead of postulating a hundred-mile-thick

Fig 5 The Earth as a hollow globe, according to the Society for Geocosmical Research

terrestrial crust, with nothing at all outside it, they preferred to believe in an Earth extending infinitely in all directions beneath our feet. Neither did they believe in the hidden central sun; they accepted our familiar Sun as being the genuine article, though they agreed that it must be half brilliant and half dark. The Sun itself did not move, or even rotate; but the inside of the Earth span round, completing one journey in 24 hours and giving us the alternation of day and night.

In their various publications, Herr Schmidt and his colleagues confirm that the stars are spots on a crystal sphere surrounding the Sun. Australia cannot be seen overhead because of the brilliance of the sky, and also because light does not travel in straight lines; it follows the curvature of the inside of the Earth. The last paragraph of a booklet published in 1957 by Herr Müller-Murnau is instructive: "Mankind can only gain if it recognizes that for good or bad it is living in an enclosed space, and at the same time is embedded in the cosmos."

Over the years I have had some correspondence with Herr Schmidt, and at one stage I put him in touch with the Flat Earth Society. It seemed an excellent idea to bring them together, but in the event they did not get on really well, and the final outcome was inconclusive. For some reason they did not seem able to appreciate each other's point of view.

So much, then, for Koresh, Magdeburg and Herr Schmidt —and it is perhaps time to turn back to an even older theory, that of the solid sky. As we have noted, all the ancients believed in it. In modern times we have one of our major Independent Thinkers, Mme. Gabrielle Henriet, who is a strong supporter of the idea.

Mme. Henriet believes that the Earth is not freely suspended in space, but is resting on the floor of a cavity whose walls surround it on all sides. The vault is rather steep, and is shaped like a pyramid; it is made up of bright metallic material, and we know something about its composition, because the meteorites which fall to the ground and can be studied in our museums are simply fragments of the vault which have been knocked off (literally, not metaphorically). Mme. Henriet can give proof of this, since in her book she states that she has actually seen the vault with the naked eye. Periodically it is illuminated by lightning flashes during thunderstorms, and can be glimpsed for several consecutive seconds.

There is another proof, also. The Moon and planets are not solid bodies, but are merely luminous, transparent disks. What we normally believe to be the mountains and craters of the Moon are nothing more nor less than "the topographical features of the solid vault of the sky, which are illuminated and thrown into relief by the luminous and transparent disks" which we call the Moon, Mars, Jupiter and so on. Beyond the solid vault is a tremendous mass of water, and rain is due to this water seeping through temporary cracks in the vault.

Fascinating indeed. And, after reading the foregoing, I am sure you will understand my disappointment when Mme. Henriet, for the perfectly valid reason already stated, decided that it would not be possible for her to join me on television.

And yet all these theories, remarkable though they are, pale before the intellectual arguments of Mr. John Bradbury—lecturer, broadcaster and scientific revolutionary in the best sense of the term. To him we must now turn.

4

THE BRADBURY UNIVERSE

In the autumn of 1966 a particularly interesting meeting was arranged by the Cambridge University Astronomical Society, England. The speaker was Mr. John Bradbury, who lived (and still lives) at Ashton-under-Lyne. Mr. Bradbury is not a professional astronomer; he is in fact a chiropodist, but for many years he has been engaged in scientific research, and his findings are, to say the least of it, remarkable. He has developed an entirely new form of mathematics; he has constructed a revolutionary telescope, with which he has been able to observe the stony metallic casing of the universe; and he has demonstrated that the Moon is a small body covered with plasticine phosphorus.

Mr. Bradbury's lecture was very well received. He spoke for an hour, and then answered questions with calm confidence. It is probably true to say that the professional astronomers in the audience were at variance with some of his ideas—as when he explained how light travels at nil velocity through the sub-semi-vacuum inside the hollow metallic boundary of the universe; but the applause at the end of the meeting was loud and prolonged. Since then, Mr. Bradbury has lectured at other universities also, and he has made various broadcasts both in sound radio and on television.

I was delighted when he was able to join me in what was, I think, a fruitful discussion. Mr. Bradbury has so many theories that to describe even half of them would take many hours. Therefore, I will do my best to give a summary of the points which he had brought out in his broadcasts and his University lectures.

Mr. Bradbury, as a practical experimenter, is well aware that theories cannot be built up unless they rest upon some kind of observational evidence. This means building a telescope—and an ordinary one will not do; it must be of special design. The Bradbury solution is to use as

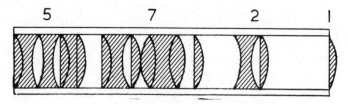

Fig 6   Mr. Bradbury's telescope

many lenses as possible, which means, of course, that the amount of light collected will be increased; it is logical to assume that two lenses will collect twice as much light as only one. The lenses need not be large, or of high-quality glass. For his latest telescope Mr. Bradbury has used fifteen lenses, of approximate diameter two inches each. These are put into a tube, as shown in the drawing.

Now, each time you add a lens, the shape of the object under observation is altered. The brilliant planet Venus brings this out excellently, and in the fifth shape it appears as a cross. More importantly, the telescope is able to show the actual background casing of the universe. There are only two major difficulties with a telescope of this kind. First, the instrument is not easy to use; one has to look through a very small pinhole, and I admit that I personally found it tricky, though this may be because I am so used to looking through my own astronomical telescope—which is constructed upon the conventional rather than the Bradbury pattern. Secondly, the telescope will show nothing which can be seen with the naked eye, which means that lining it up with a distant object is bound to be rather arbitrary.

Mr. Bradbury states that with ten lenses, the instrument has become powerful enough to show the vacuum which lies high above the Earth. Below this is a semi-vacuum, in which the celestial bodies move, and below this again is what he calls the sub-semi-vacuum. Actually, this cosmological view of the universe is rather difficult to put into words, but Mr. Bradbury's scheme of it, given here,

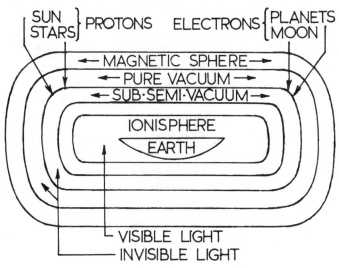

Fig 7 The Universe, according to Mr. Bradbury

should explain what is meant. Note that the outer casing is magnetic—a point which is important, and which must be borne carefully in mind.

Before we go any further, let us look more closely at the shape of the Earth. The Flat Earth Society's views have already been described, but the Bradbury version is somewhat different, since the Earth has a flat top and a convex bottom; the North Pole lies in the middle, and so, in fact, all directions are south. To talk of "north", "east" or "west" is meaningless. One cannot go as far as the outer edge of the Earth, because anyone who tries to do so will be gently coerced back toward the middle by the unobtrusive but omnipotent effects of magnetism.

Now for some proof. Mr. Bradbury maintains that the apparent curvature of the horizon is an illusion, and I can do no better than quote his own words:

"Take two examples. First, here is a man who lies down in a field and looks up into the sky; the sky then seems the shape of an umbrella. If you get up and walk for some distance, the same thing is visible; you seem to have brought the umbrella-shaped sky with you. Next, consider someone sitting on a deck-chair on the sands looking out to sea. He will see an apparently curved horizon; but the convexity is due to the real convexity of the human eye. The sea, as you will find, always appears at eye level... The sky is solid, and, as has been proved, the background of the univere is solid metallic stone."

Nothing could be fairer than this. Mr. Bradbury also points out that as one goes up, the temperature falls (this is the well-known meteorological lapse-rate); and if one goes sufficiently high, the temperature must fall below that at which air will liquefy. Therefore, at sufficient height— perhaps fifty miles—we must reach a region in which there is intensely cold, liquid air. Above this lies the sub-semi-vacuum which we discussed earlier.

What, then, of the Sun and the Moon? In the Bradbury

universe, the Moon is a mere 220 miles away, so that with the special telescope its apparent distance can be reduced to about forty yards. The Moon is not perfectly flat; it is slightly convex, and is made of carbon. The lunar phases are due not to the changing illumination of sunlight, but to something much more radical. Mr. Bradbury explains that as the carbon-disk Moon moves in the sub-semi-vacuum, it picks up some material such as plasticine phosphorus, which is circulating above the liquid-air zone. When waxing, the Moon collects this material; when waning, it sheds it, and this cycle occurs regularly. Once, in 1953, Mr. Bradbury's telescope showed him a fingerlike projection from the Moon (see frontispiece), so that evidently the Moon had collected more plasticine than usual. Unfortunately an ordinary astronomical telescope would not show the projection, and nobody else was able to confirm it. I asked Mr. Bradbury whether he could be sure that there was only one fingerlike projection and not two, but he was adamant on this point.

The Sun, at a distance of at least 400 miles, is further away than the Moon. It cannot be hot, because if it were then we would clearly be unable to receive any heatwaves. Instead, it produces invisible rays which come through to us by way of the sub-semi-vacuum. If actual heat were transferred through the upper belt of liquid air, then of course the air would be melted, and the Earth would experience continuous, steady rainfall—which does not happen.

Day and night can be explained by the fact that part of the upper casing acts as a polaroid, and regular turning of this material produces alternate transparency and opacity. Light moves instantaneously; therefore it can be said to have nil velocity, and does not 'travel" in the accepted sense of the term. According to Mr. Bradbury, it is made up of three conditions of matter, and our accepted ideas of "colour" are all wrong. This can easily be demonstrated.

As he says: "Red light is actually green. You can prove this by plugging in an electric fire. You get the green light first, and then the red, which are of course interchangeable. Green is the only true colour, and includes all the other so-called colours of light."

I particularly admire Mr. Bradbury's definition of "light". I asked him what it was, and he replied, simply: "Light is darkness, lit up." Nobody, however orthodox, is likely to quarrel with this.

Tides also have come under scrutiny. These are presumably due to mercury mines in Australia. When the Sun goes below the Earth (still, of course, keeping within the pure vacuum) the mercury is heated; this deforms the Earth, and the ocean tides result.

Since Mr. Bradbury first proposed his theories, men have landed on the Moon—or have they? If the Moon is an extremely small body, covered with plasticine phosphorus and made of carbon only an inch or two thick, it would be rather difficult to land there. There can be only two possible explanations. Either the Bradbury picture is wrong, or else the Apollo astronauts have not reached the Moon at all. Mr. Bradbury favours the latter explanation. He considers that instead of going straight up for a distance of a quarter of a million miles, the Apollo space-craft went sideways, diverted by the force of the magnetic outer casing; had the vehicles continued in a vertical direction they would have met the layer of liquid air, battered their way into the sub-semi-vacuum, whizzed through the pure vacuum, and probably smashed themselves to pieces on the outer metallic casing, producing a sad shower of meteors. He maintains that instead of going to the Moon, the astronauts have actually landed in Tibet. This is a very lofty area, which accounts for the lack of thick atmosphere as well as the absence of life: moreover, the political situation in the East at the moment would preclude any announcement of such a visitation. The return journey to splash-down in the ocean would also be sideways.

Such are Mr. Bradbury's theories. What can one make of them?

That they are wildly unconventional is only too clear. They bear no relation whatsoever to anything in accepted science. Mr. Bradbury himself is well aware of this, and he does not mind in the least. He appreciates, too, that because there is no common factor between his universe and everybody else's, there is no real way in which a discussion can "get off the ground". Unlike some of those who put forward revolutionary ideas, he has no feeling at all of being persecuted or misrepresented; in outlook he is a true pioneer, and his aim is to work away, quietly and patiently, putting forward his theories, building his apparatus, and waiting for the time when his cosmology will replace that of Newton and Einstein. At the moment he is beginning serious experiments in photography, using small lenses coated with opaque substances. I have no doubt that the results will be quite fascinating.

I do not personally agree with Mr. Bradbury's ideas. I can hardly be expected to. But I have the greatest personal admiration for him as a man who has had the moral courage to throw overboard every semblance of orthodoxy, and strike out on his own, without fear of any ridicule or scorn which he may draw on his head. As he has shown in his University lectures and in his broadcasts and television appearances, he is ready to go upon his way, absorbing every scientific advance and putting his own interpretation on it. Society would be the poorer without men such as Mr. Bradbury. To me, he is the supreme example of the Independent Thinker.

5

THE FRIGID SUN

In the history of astronomy there have been some truly great names. All the outstanding personalities have been Independent Thinkers in every sense of the term, and some of them have held views which are not generally accepted today. One is reminded of Johannes Kepler's "five regular solids", and we must also remember that Sir Isaac Newton himself was an astrologer-cum-alchemist when he was not making revolutionary mathematical advances or compiling the Laws of Gravitation.

Coming to more modern times, we have Sir William Herschel, who became the most famous astronomer of his day. He was still President of the Royal Astronomical Society at the time of his death in 1822, and as a pure observer he has probably never been surpassed. He made his own telescopes, which were unrivalled; in 1781 he achieved lasting fame by his discovery of a new planet, the remote world we now call Uranus. Later, he drew up the first reasonably correct picture of the way in which our star-system is shaped. To list all his contributions and discoveries would take many pages; suffice to say that they were unmatched. He was never Astronomer Royal, but he was given the official title of King's Astronomer, plus a royal annuity which allowed him to give up his career as an organist and devote his whole life to astronomy. In most spheres Herschel's authority was supreme, but in one important issue he was a rebel. Unlike most of his contemporaries, he believed that all the planets were inhabited, and that the existence of intelligent beings on the Moon was "an absolute certainty". Moreover, he was convinced that beneath the hot surface of the Sun there was a cool, pleasant region where people lived.

Modern astronomers could not be in greater disagreement. To them (and, let me admit, to me) the Sun is a star; that is to say, a globe of incandescent gas 865,000 miles in diameter, with a surface temperature of 6,000 degrees Centigrade and a temperature near the core of something like 14 million degrees Centigrade. The Sun is not "burning" in the usual sense of the word. Deep inside it, hydrogen atoms are combining to make up atoms of another element, helium; each time this happens, a little mass is lost and a little energy is released. At the cost of losing a grand total of 4 million tons every second, the Sun continues to radiate—as it has done for the past 5,000 million years, and as it will continue to do for at least 6,000 million years in the future before its supply of hydrogen "fuel" begins to run low. All other solar-type stars are behaving in the same way, though admittedly the very young and the very old stars have rather different methods of doing things.

However, all this is by the by. To recapitulate: Herschel believed that the Sun must be cool inside, but he did not doubt that the actual surface is very hot indeed. At that time, of course, nobody really knew how the Sun and other stars functioned; it was only just before the outbreak of the last war that astronomers made up their minds. By then, the idea of Sun-men (or Solarians, to use a more attractive term) had long since been officially given up, and so had the notion of a cool interior—except by a few Independent Thinkers, to whom we must now turn. It has been maintained that the Sun is not hot at all, even at the surface, and that it may not be a solid body, in which case a well-aimed space-ship would go straight through it.

So far as I have been able to find out, the latter notion was first proposed in or about 1947 by an Argentinian lawyer named Navarro. To him, the Sun was simply the focal point of rather mysterious particles which came together at the centre of motion of the Solar System. He added, for good measure, that the Earth is hollow and that people live on the inside of it as well as on the surface (shades of Captain Symmes). Some years later, in 1952, the matter was taken up by another lawyer, this time a German named Godfried Büren, who reverted to the age-old theories of Herschel, and stated categorically that the Sun had a cool inner globe covered with vegetation. The dark patches on the brilliant surface, misleadingly known as sunspots, are merely vents through which the dark core can be seen. Büren went so far as to offer a prize of 25,000 German marks (around £2,000 or $4,800) to anyone who could disprove his ideas. What followed was as unexpected as it was ironical. The leading astronomical society in West Germany accepted the challenge, and put forward a whole string of objections to the picture of a temperate Sun. Rather naturally, they considered that their arguments were conclusive. Herr Büren did not, and refused to pay up. The result was a complicated court case, and to Herr Büren's amazement the verdict went against him. I have never been able to discover whether the 25,000-mark wager was actually paid!

More recently still we have, of course, the universe as

outlined by Mr. Bradbury, to whom the Sun is a small, cool object at a distance of about 400 miles. But as we have seen—and as Mr. Bradbury is the first to admit—his theories are so far removed from orthodoxy that one really has either to accept them *in toto* or else reject them in equal *toto*. Quite different is the attitude of the Rev. P. H. Francis, the Vicar of Stoughton in Sussex, England only a few miles away from my own home in Selsey. Mr. Francis is emphatic that the Sun is not hot. To quote from his fascinating book, *The Temperate Sun*: "The popular notion that the Sun is on fire is rubbish, and merely a hoary superstition, on a par with a belief in a flat earth, an earth resting on the back of a tortoise or an elephant, or a sun revolving round a stationary earth...It rests on no sure basis of evidence; and if it is discarded, great simplifications become possible in the science of astronomy, geology and physics, and many other branches of science can be placed on surer foundations."

Strong words—but Mr. Francis has evidence to back them up. And let it be said at the outset that he is no amateur. He has an extremely good M.A. degree in Mathematics from the University of Cambridge; he has taught the subject in schools for many years, with great success; and he is an experienced lecturer and writer. He is also very active in the affairs of his parish, and is as popular as he is energetic. I have talked to several of his ex-pupils, and all pay testimony to his ability as a teacher as well as to him personally.

It is important to stress this, because—and I say this without any intention of being caustic—many of the Independent Thinkers of today are people who stay well outside the boundaries of accepted science, and do not have what may be called scientific backgrounds. Therefore Mr. Francis is, if not unique, at least unusual. Mathematics degrees from Cambridge are not given lightly, as any graduate knows. I would very much like to have such a degree myself, but I know my limitations!

Armed with this introductory knowledge, then, let us see what Mr. Francis has to tell us.

During a television programme in which I took part, I went to see him in his delightful country vicarage, together with our technicians. It was a most enjoyable afternoon, and the Sun shone gaily down, giving at least the impression that it was warming us. True, there was a temporary break when we accidentally set the Vicarage on fire. To add to the visual effects, it was suggested that we might add a little paraffin to the fire which was burning in the drawing-room hearth. Unfortunately, we overdid things somewhat. During the interview, when the television cameras were switched on, we became conscious of a dull roaring noise, which steadily became more and more obtrusive. It was followed by a shower of sparks, and clouds of black smoke began to belch from the chimney. Prompt and decisive action by all concerned brought the situation under control, and the only real damage was to the carpet, but it was quite a hectic few minutes. Typically, Mr. Francis was not in the least annoyed. He said—with truth—that it had been a most interesting afternoon!

"The notion of an incandescent Sun is childish, and not worthy of a grown-up person; it is almost incredible that people who call themselves scientists should not have thought of questioning this primitive belief." So wrote Mr. Francis in the journal *Astronomy Today*, in 1969. But, of course, though the Sun does not send heat to the Earth, it *causes* heat on the Earth, which is a very different thing. To show what is meant, Mr. Francis explained to me that the best analogy was that of an electric generating station. When one switches on an electric fire, heat is caused, and the room becomes pleasantly warm even when an Arctic-type blizzard is raging outside. And yet the generating station itself is not on fire; it is not even hot —and one can go up to it without any feeling of discomfort. Heat is caused, but not sent direct from a central hot body.

The argument is devastatingly simple. There is another point also, brought out by Mr. Francis during our talk and also in his book. An ordinary household vacuum-flask is made up of a central container, outside which is an outer casing; between the two there is vacuum (hence the name). Now, heat cannot cross a vacuum. When you fill the flask with hot tea or coffee (or, if you happen to be Japanese, hot *sake*), the heat cannot escape, because it cannot cross the few hundredths of an inch of vacuum and reach the outer case. Therefore, the liquid stays hot. There are 93 million miles of vacuum between the Earth and the Sun, so that, quite obviously, no direct solar heat could ever get to us.

Moreover, there is the contention that the Sun would be unable to send heat out into empty space. Space is nothingness, and even gravitation can affect only the solid material bodies which exist. As Mr. Francis says, "Nature would not be so foolish as to exert gravitational forces on things that don't exist," and neither is it at all likely that heat would be sent out aimlessly into the void of space.

As an observational proof, Mr. Francis takes the circular outline of the Sun's disk as seen from Earth. If the Sun were really a molten mass of seething gas, this regular outline would often be distorted. We must also bear in mind that temperatures on our world are relatively steady, which could hardly be the case if the heat were sent to us from a raging, uncertain Sun. "If the Sun is a hot body, it is improbable that life on Earth will exist tomorrow."

Since it is very likely that life here will still go on tomorrow (at least, I hope so), we are entitled to ask Mr. Francis exactly how the Earth is kept warm. Mr. Francis explains that this is due to the fact that it has its own self-contained energy cycle, and this energy continuously changes form because of the altering relative positions of the Sun and the Earth. Quite possibly the most important factor is the Sun's electrical charge. This, however, does not mean that electrical energy is actually sent from the Sun to the Earth. What happens is that the Sun's charge excites or induces electrical actions within the atmosphere of the Earth, and it is these actions which produce heat. Another simple natural phenomenon can be accounted for in this way. If you climb a mountain, the temperature falls; if you happen to ascend to the tip of (say) Mount Everest or Popacatapetl, you will have to take your winter woollies. According to Mr. Francis, this is because the atmosphere is densest at low levels, so that the Sun can act upon the greater amount of energy there and so produce more heat.

Next, we may ask why the Earth does not lose heat by radiation into space. The answer is that it can't; and again we come back to that significant device, the humble vacuum-flask. Mr. Francis begins by pointing out that a polished or silvered kettle is not so easily heated as a black one. (If you doubt this, try it out on the kitchen stove; the effect is very marked.) To prevent heat radiating away from the inside of the vacuum-flask, the cunning makers coat the walls of the flask with a silvery substance. Now, the light produced by the Sun on the Earth makes the Earth shine, and this prevents radiation—simply because the daylight "silvers" those parts of the Earth which are turned sunward. Similarly, that part of the Moon which faces the Sun becomes bright and silvered. We must also bring in the earthshine, that is to say the faint luminosity of the "dark" side of the crescent Moon; it is visible every month, and was correctly explained many years ago by the great painter-cum-scientist, Leonardo da Vinci. The admittedly small amount of earthshine is sufficient to stop the heat sent to the Moon from the Earth being radiated back towards the Earth and away from the Moon. "Each heavenly body becomes silvered, or shines, to a degree corresponding to the amount of light produced on it; and this prevents radiation from it."

Once the revolutionary idea of a temperate Sun has been accepted, everything else falls neatly into place, and we can proceed with our investigation of the universe. Mr. Francis' interpretation differs widely from that of either Mr. Shenton or Mr. Bradbury, mainly because he is a trained and highly-qualified mathematician; and in his book *The Mathematics of Infinity* he presents an impressive case. His first contention is that the Earth lies at the centre of the universe, and can never move away from it. This is not a reversion to the age-old ideas of the Greeks; it is more nearly allied to Einstein and the theory of relativity. To quote Mr. Francis: 'Any body in space is at the middle point of any straight line drawn through it. Two heavenly bodies could move relatively to each other along a line drawn through them, but each would remain half-way between opposite infinite points of the line. Hence any body in space is at the centre of the universe, and can never move away from that position. If the universe were finite, absolute motion would be possible, because it would be possible to move nearer to its furthest points; but since it extends to infinite distances in all directions, it is not possible to approach nearer to its limits, and all bodies must remain half-way between infinitely opposite points of the universe—that is to say, at the centre of the universe."

All this is perfectly clear, but it must be remembered that infinity lies at different distances from us according to the direction in which we look. As Mr. Francis told me, infinity may be only a few inches away in some particular direction, while in another it may be millions of light-years. It surrounds the Earth, but it is not regular. Infinity has a curved reflecting surface, and this leads us on to an explanation of the things we call "stars". Having rejected the notion that the Sun and the stars are huge bonfires, we must cast our net elsewhere.

If the surface of infinity surrounding the Sun (and the Earth, which also occupies the exact centre of the universe) were regular, then there would be uniform illumination, and the whole sky would appear light and featureless. This does not happen, because the reflecting surface of infinity is very far from being regular. We can draw an analogy with a candle in a "hall of mirrors"; the light will be reflected from one mirror to another, and will end by producing an infinite number of images. It is the same with the stars. At infinity, the Sun's image appears as a point. This strikes the curved surface of infinity, producing a star; the image is reflected on to other parts of the curved surface, and more and more stars are produced. I have made an attempt to draw what happens, but my sketch is not likely to be very accurate, since to work out the precise profile of infinity is not an easy matter.

This is not all! To speak about distances greater than infinity is absolute nonsense. Infinity is the limit of the universal background, and nothing can be greater than that. However, light from the Sun can pass through infinity; it then spirals round the universe and reappears at a totally different point in the sky, albeit with its brilliancy much reduced. Very faint stars are therefore reflections of the back of the Sun. There is no system apart from the Solar System, and the planets are luminous bodies which produce their own images on the curved reflecting surface of infinity. To sum up in Mr. Francis' words: "No star has a real existence, any more than the image of a candle in a mirror has a real existence."

Nothing could be further removed from the conventional picture of a star as a vast, immensely hot body, pouring radiation into space, and perhaps sending out as much energy as ten thousand Suns put together. We are back to a much more placid universe, which contains relatively few bodies, and in which no object really interferes with any other. The Sun has reverted to its position of supreme importance, not because it is hot, but because it is necessary for the impetus for the Earth's life-cycle to operate.

Mr. Francis first published his theories some years ago. Since then, the Space Age has begun. Men have flown above the atmosphere, and have even landed on the surface of the Moon (the silvered surface, of course; nobody has yet braved the rigours of a lunar night). Rockets have been sent as far as Venus and Mars. Obviously the whole process is a risky one, as Apollo 13 showed only too clearly; but Mr. Francis points out that one aspect of the danger will be removed if he is right and the astronomers are wrong.

If a space-ship goes off course, he says, there is always the likelihood that it will be drawn into the Sun. With a sur-

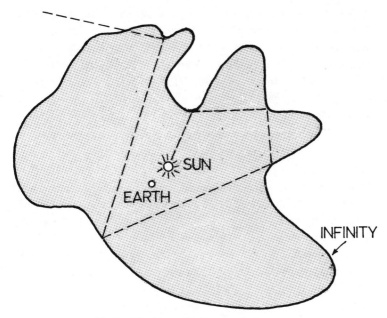

Fig 8 Mr. Francis' theory of the stars

face temperature of 6,000 degrees, this would be rather unfortunate for the occupants, who would be fried like eggs long before they hit the outer gases and disappeared with a slight sizzle and a puff of smoke. But if the Sun is a cool, temperate body, of the same basic type as the Earth, there can be no danger from it. It might be possible to land there, and indeed Mr. Francis believes that it would be a particularly suitable place to visit.

How can we find out? One possibility, of course, is to send an unmanned rocket probe there, and see whether the probe (a) is burned up, (b) passes right through the place where the Sun seems to be, or (c) achieves a soft landing, as the Luna and Surveyor vehicles of the mid-1960s did on the Moon long before the first human passengers went there, and as the Russians have since managed to do with Venus and Mars. No doubt a solar probe will be launched in the foreseeable future.

Summing up: we have an Earth which, like everything else, lies in the exact centre of the universe, and is motionless from a relativistic point of view. Heat is produced by the Earth's self-sufficient energy cycle, initiated by energy (not warmth) from the Sun. The planets are luminous bodies; the stars are simply reflections of the Sun on the curved surface of infinity. The Sun is no danger to spacetravellers, and neither can we on Earth be either fried or frozen in the future, because we depend upon our world alone. Silvering of the sunlit hemisphere effectively prevents radiation away into space.

But the most fascinating prospect of all is that of landing on the solar surface. Let me be honest: I greatly fear that Mr. Francis is wrong, and that no such thing will ever be possible even for one of the tough, grizzled spacecaptains of the twenty-first century. But I hope he is right —and I would give a great deal to see the first intrepid astronaut step out and gaze around, in wonder and awe, at the green fields of the temperate Sun.

6

DR. VELIKOVSKY'S COMET

Of all the Independent Thinkers of modern times, none has caused more controversy than Dr. Immanuel Velikovsky. His only possible rival is Mr. George Adamski, of whom more anon. I did know Adamski, and even interviewed him during one of his infrequent visits to England. To my great regret, I have never met Dr. Velikovsky. Though he is of Russian birth (he was born at Vitebsk, in 1895) he has lived for many years in the United States, where he carries out the researches which have made him so famous.

Briefly, his theory is that the planet Venus used to be a comet—and that in Biblical times it passed close to the Earth more than once, causing tremendous global upheavals, and behaving rather in the manner of a cosmical ping-pong ball. This in itself is extremely revolutionary, and Dr. Velikovsky's original book, *Worlds in Collision*, caused raised eyebrows when it first burst upon the scientific world in 1950. But before discussing it, let us hark back a few centuries and consider Dr. William Whiston, a contemporary of no less a person than Sir Isaac Newton.

Whiston was a scholar and clergyman, who actually succeeded Newton as Lucasian professor of mathematics at Cambridge, England. This was in 1710. Fourteen years earlier he had published a learned tome, *A New Theory of the Earth*, in which he tried to correlate the events of the Book of Genesis with the wanderings of a comet. More than this, he believed that the Earth itself used to be a comet in bygone ages, and that it moved in a much more elliptical path than it does now. Each time it passed close to the Sun it was violently heated, and this explains why even today the Earth is hot near its centre (modern estimates put the core temperature at several thousands of degrees). Gradually the Earth's orbit became more circular; the regular overheating stopped; and the comet turned into a planet, after which Man made his entry upon the terrestrial scene.

Unfortunately, the first men were far from blameless in their outlook, and they sinned grievously. The result was that a small comet was sent to by-pass the Earth, giving it a movement of rotation. A huge tide swept over the lands (this, of course, was the Biblical Flood), and the watery particles from the comet produced a rainstorm lasting forty days and forty nights, which today would be something of a meteorological record, even in Oldham. Whiston gave the date of this disaster as either November 28, B.C. 2349, or December 2, B.C. 2926—he was not sure which. He also predicted that the comet will eventually come back, so that the Earth will once again be violently perturbed; it will resume its cometary nature, and will cease to be habitable.

Whiston's ideas were not well received in scientific circles, and his whole career was rather stormy; I mention it here because of his belief that comets can change into planets and vice versa—a belief also held by Dr. Velikovsky, whose whole thesis is indeed based upon it.

Let it be said at once that Velikovsky's doctorate is perfectly conventional. For some time he practised as a G.P. in Jerusalem. He then trained as a psychoanalyst, and practised at Haifa and Tel Aviv. It was not until 1939 that he went to America, where he has remained ever since. He is not a trained astronomer or mathematician, but psychoanalysts and psychiatrists in general are not to be classed with other men, and Dr. Velikovsky is no exception. It may be unkind to suggest, as one critic did, that his methods of historical research are similar to those of the man who identified Moses with Middlebury by the simple expedient of deleting the -oses and adding the -iddlebury; but let us look more carefully at the results of his labours.

All his three books are heavily annotated, and every Biblical reference is correct. In fact, Dr. Velikovsky has done his homework extremely well in this respect. It is his scientific reasoning which has caused his contemporaries to frown thoughtfully or, in some cases, to explode in anger.

He begins, fittingly enough, with the giant planet Jupiter, which—he says—suffered a tremendous outburst, and shot out a comet which later became the planet Venus. It is difficult to tell just when this happened, but cosmically speaking it was not very long ago; to Dr. Velikovsky, astronomical time is to be measured in thousands of years rather than in millions. Originally the comet Venus had an elliptical path, and this led it into a series of strange encounters. It first passed near the Earth in B.C. 1500, at the time when the Israelite Exodus was being led by Middlebury (apologies: by Moses). The result was that the Earth temporarily stopped spinning, or, at least, slowed down; and the Red Sea was left high and dry for long enough to allow the Israelites to cross. Then, however, came huge upheavals, as the Earth's surface twisted and turned under the influence of the gravitational pull of the comet Venus. Petrol rained down—and, incidentally, our modern motor-cars are powered by precisely this petrol, "remnants of the

intruding star which poured forth fire and sticky vapour".
Modern Governments have therefore cause to be grateful
to the wandering Venus; but for it, there would be no
petrol for them to tax.

Conveniently, the Earth's rotation started up again just
in time to swallow up the Egyptians who were in hot pur-
suit of the retreating Moses. But this was not all. Having
made its presence felt, the comet Venus withdrew into the
limbo, but came back for a second visit two months later
so that it could produce the lightning, thunder and other
spectacular effects noted when Middlebury was given the
Ten Commandments on Mount Sinai. Subsequently, some
of the materials in the comet's tail rained down on Earth
in the form of manna upon which the Israelites fed for the
next forty years.

One might have thought that this would have been
enough. But no! The comet Venus continued its erratic
career, coming back to see us several times—once, for instance, to shake down the walls of Jericho. However, the
day of reckoning was at hand. The comet collided with
Mars, and had its tail chopped off, so that it stopped being
a comet and started being a planet. Not to be outdone,
Mars itself moved closer to the Earth, and indeed nearly
scored a bull's-eye in the year B.C. 687. Further encounters
took place, and Dr. Velikovsky links various Old Testa-
ment disasters with those approaches of Mars. On one
occasion the Earth turned head over heels, so that for a
time the Sun rose in the west and set in the east. Neither
did the Moon escape unscathed; both its distance and its
orbit were frequently changed. Then, gradually, the situation eased. Mars retreated again, and resumed its former
path; Venus, no longer a comet, settled down into a peaceful and almost circular orbit; the earthquakes, tidal waves,
floods, cyclones and other terrestrial events subsided.

All this is fascinating, but there are some objections
which spring to the mind at once. For instance, orthodox

scientists (such as myself) are bothered by the attempted identity between a comet and a planet. To us, a comet is a flimsy body, made up of small particles—mainly ices—together with extremely thin gas. It has even been said that a comet is "the nearest approach to nothing that can still be anything", and its mass, compared with that of a planet, is absolutely negligible. There have been observed cases of cometary encounters—notably in 1770, when a comet known by the name of its discoverer, Lexell, passed so close to Jupiter that it invaded that planet's system of satellites. Neither Jupiter nor his family seemed to be in the least perturbed, but the comet had its orbit completely altered. And on at least two occasions in the moderately recent past the Earth has passed through the tail of a large comet without coming to the slightest harm. Represent the mass of the Earth by that of an elephant, and the mass of a comet will be approximately represented by an ant. Obviously, it would take a great number of ants to pull a charging elephant from its chosen path (even if the experiment could be tried, which I frankly doubt).

Therefore, say the anti-Velikovskyites, it is rather difficult to see how a comet could change into a planet. One might as well expect a cloud to change into a block of lead. Moreover, it is usually thought that the fossil record of the Earth goes back a great deal further than Biblical times; and there are, to put it mildly, certain mathematical obstacles in the way of supposing that the Earth could stop rotating, or that a planet such as Venus or Mars could bounce wildly about the Solar System. Minor details of this sort do not bother Dr. Velikovsky in the least. Why should they?

Quite clearly Dr. Velikovsky's theories are no more unorthodox than those of Mr. Bradbury or Mme. Henriet, but the reception given to them was entirely different. When his first book came out, in 1950, it produced a storm just as violent as those roused so long ago by the comet Venus. Technical magazines denounced it; highly distinguished scientists wrote scathing reviews of it. Macmillan & Co., the original publishers, found that universities and colleges were ready to boycott their entire book list so long as Velikovsky's *Worlds in Collision* remained on it, and after a while the publication rights were transferred to another firm. Against this, many people were impressed by the wealth of detail and by the Biblical and folk-lore references, all of which, as we have noted, are entirely accurate. Among the book's supporters were J. J. O'Neill, science editor of the *New York Herald Tribune*, and Dr. Atwater, chairman and curator of the Hayden Planetarium in New York City. To be fair, neither of these gentlemen accepted Velikovskyism hook, line and sinker; but they did believe that the book deserved serious examination and criticism. Also cited was another near-modern comet collision enthusiast, Ignatius Donnelly, whom we will meet again when we come to consider Atlantis, and who had at one time been a candidate for the Presidency of the United States.

Meanwhile, Dr. Velikovsky had been preparing more books, and they duly appeared. I do not propose to discuss them here simply because they add little to his basic hypotheses. Ever since those first hectic days, he has been citing cases in which his prophecies have been proved correct. On several occasions he did hit the proverbial nail squarely upon its equally proverbial head. He forecast the high surface temperature of Venus (the present planet, that is to say; not Middlebury's Comet). Since Venus is, on average, 26 million miles closer to the Sun than we are, and has a dense atmosphere which acts in the manner of a greenhouse, this was perhaps a fairly safe prediction; but he also forecast that the axial rotation would be slow and retrograde—that is to say, east to west instead of west to east. The latest measurements confirm that Venus really does spin in this peculiar way.

Dr. Velikovsky also forecast an excess of hydrocarbons on Venus. Hydrocarbons are members of the petrol family, and later on Professor Fred Hoyle, one of the world's leading astronomers, agreed that Venus could be rich in them, though his reasons were very different from Velikovsky's. In a pro-Velikovsky circular distributed in July 1971 from the University of Texas, much was made of this. Unfortunately, a few weeks later new results obtained from the Russian probe Venera 7, which made a gentle landing on the planet, indicated a total lack of anything of the sort. Dr. Velikovsky predicted that the atmosphere of the planet Mars was likely to consist of argon, neon and nitrogen; actually it has turned out to be almost pure carbon dioxide. When we come to moonquakes, we must be somewhat cautious. In 1969 Dr. Velikovsky had stated that "moonquakes must be so numerous that there is a bit of a chance that during their few hours on the Moon the astronauts may experience a quake". Now, minor tremors on the lunar surface do occur; this is a subject which I have been investigating myself for a long time—but it all depends upon what one means by a moonquake! The tremors are so slight that they could certainly not be felt by anyone standing there. In passing, Dr. Velikovsky believes that the mountains and craters of the Moon are less than three thousand years old.

It is, I think, only right to point out that anyone who makes a habit of coincidence-hunting will almost certainly be able to find what he wants. The Bible is extremely lengthy, and the amount of available folk-lore is almost limitless in many countries and in every possible language. It has also been maintained that anyone who writes enough words, and makes enough forecasts, will be bound to make some correct statements eventually. As was once said by a learned judge, "it is impossible always to be wrong" (though some modern politicians have made gallant attempts at it). And if the wandering comet can be assumed to return at irregular intervals, it can be held responsible for almost every catastrophe which occurred before the start of reliable historical records.

It is interesting to speculate as to why Dr. Velikovsky's works have aroused such bitter hostility. As we have seen, they are no more unorthodox than those of, say, Mr. Bradbury—who has aroused no fierce criticism at all, and whose lectures at Cambridge University and elsewhere have been received with marked, albeit stunned, fascination. There is, however, an essential difference. The Bradbury universe is so utterly removed from any official science that nobody can really criticize it; there is no common basis upon which to plant one's flag, as Mr. Bradbury himself is the first to accept. But Dr. Velikovsky is a scientist inasmuch as he is a qualified medical practitioner; and his critics were infuriated by the fact that his first book was not only brought out by an extremely well-known and reputable publisher, but was put out as an important new concept of the history of the Solar System. The obvious scholarship in it obscured the fact that Dr. Velikovsky is neither a physicist, an astronomer nor a mathematician, and is therefore no more academically qualified in this line of research than any of the other Independent Thinkers whom we have been discussing in these pages.

I have been unable to find out just what Dr. Velikovsky thinks will happen in the future. At the present time Venus and Mars are settled in their orbits; but can it be that in the fullness of time Venus will awake once more, turn back into a comet, and set off upon a new blaze of destruction, producing a fresh, terrifying tail and resuming its wild waltz in the Solar System? Shall we find that the Earth has turned turtle yet again, so that for a change the Sun will rise in the north and set in, perhaps, the south-east? Will there be fresh showers of petrol and manna—and if so, will our Governments have to introduce a special rationing system to ensure fair distribution of it?

These are weighty questions indeed, and we must bide our time. Meanwhile, Dr. Velikovsky is still hard at work, utterly convinced that in the future his reputation will stand with those of Newton and Einstein, and leaving the blind, reactionary orthodox astronomers to continue their futile investigations of the comet Venus.

7

ICE, ICE EVERYWHERE

Generally speaking, Independent Thinkers of today make little impact upon the scientific policies of nations— or, for that matter, upon orthodox science itself. The Flat Earth Society has never been officially recognized; the International Astronomical Union pays scant attention to Mr. Francis' cold sun; even Dr. Velikovsky has not been discussed at Government level. However, there are one or two notable exceptions. In Russia, some years ago, the whole science of genetics was turned upside-down by one Academician Lysenko, who produced a whole crop of revolutionary theories together with innumerable experiments which, unfortunately, failed to work. But perhaps the most remarkable case of all concerns an Austrian engineer named Hans Hörbiger, whose theory of "universal ice" became so popular in Germany just before the outbreak of the war that it was almost a part of Nazi doctrine.

Herr Hörbiger's chief follower was the famous German lunar student Philipp Fauth. Broadly, the theory maintained that the key to the whole cosmos is nothing more nor less than ice. The Moon is ice-covered; so are the planets; even the so-called stars of the Milky Way are simply ice-crystals. When a block of ice spirals downward and falls into the Sun, it produces one of the darkish patches that we call sunspots. The resulting explosion blows fine ice-powder far into space, so that it finally reaches the Earth and produces high-altitude cirrus cloud.

To be precise, Herr Hörbiger was not the first to suppose that the Moon might be ice-covered. The idea was put forward in the 1880s, first by a Norwegian named Ericsson and then by a British tea-planter, S. E. Peal, who went so far as to publish a book about it. Not, of course, that either of these gentlemen believed in "universal ice". They didn't. Moreover, at that time the surface temperature of the Moon was not known, because the tiny quantity of lunar heat received on Earth is very difficult to measure. The problem was solved by an Irish nobleman, the fourth Earl of Rosse, who set up sensitive equipment at his home in Birr, County Offaly. He showed that in the middle of the long lunar day, the temperature there rises to well over 200 degrees Fahrenheit. We now know that the fourth Earl was entirely correct, and that in many ways he was scientifically several decades ahead of his time. But it was not for a surprisingly long period that his results were generally accepted, and even in the early part of our own century many astronomers still believed that the Moon's surface could never rise above freezing-point.

By 1930 the situation was becoming clearer. The Earl of Rosse had been well and truly vindicated, and a temperature near or even above that of boiling water does not seem very favourable for the existence of ice. When I first became interested in the Moon, which was around 1934, I naturally regarded the "glaciation theory" as being as outdated as a dodo. However, I had reckoned without Hörbiger's disciples, Philipp Fauth in Germany and Hans

Bellamy in England.

Hörbiger's classic book, *Glazial-Kosmogonie*, had appeared as long ago as 1913, but I had not read it, because (rather naturally) it was written in German, and my own knowledge of the Teutonic language was—and still is— limited to "Damit", "Besonders", and "Donner und blitzen". Since it has never been fully translated, I have not read the complete version even now; it runs to almost eight hundred pages, and is crammed with diagrams, photographs and portmanteau words. From all accounts of it, plus the sections which have been rendered into English, it is a mixture of mysticism and what its author sincerely believed to be cold, sober, scientific fact. Like so many Independent Thinkers (though totally unlike some of those described in this book, notably Mr. Bradbury and Mr. Francis), Herr Hörbiger was firmly convinced that he was being persecuted by Orthodoxy, and that all official scientists were his personal enemies. From his own point of view, he was unlucky not to have lived long enough to see the remarkable rise to favour of "WEL" (Welt Eis Lehre, or Cosmic Ice Theory) under the Nazis.

First, let me give a broad outline of WEL, as postulated by Hörbiger and also by Hans Bellamy rather later on (it has been modified since). We begin with space, which is filled with rarefied hydrogen. The Sun is the most important body in the universe—the stars, as we have noted, are mere chunks of ice—and is gradually drawing all the planets in toward it. This is because the planets are not moving in empty space; they have to push aside the low-density hydrogen, which sets up slight but appreciable resistance and slows them down. Very small planets can also be "braked", and will drop into the larger ones. Practically everything, apart from the Earth, is made up of ice, or is at least ice-coated. The Moon's glacial covering is 150 miles deep; that of Mars extends downward to 250 miles below the planet's surface, and the famous Martian canals are simply cracks in the upper ice-sheet.

At the moment the Moon is spiralling down towards us, and will eventually be disrupted. (This will probably happen well before the Earth itself plunges into the Sun, producing a particularly large sunspot as it is snuffed out.) But our present Moon is not the first of its kind. There have been at least six previous moons, all of which have approached the Earth quite closely before being disrupted. Each time this has happened there have been major upheavals, accounting for events such as the extinction of the dinosaurs some 70 million years ago.

According to Hörbiger, and also to Hans Bellamy in the 1930s, this penultimate Moon was at the peak of its career in near-prehistoric times. As it spiralled inward, braked remorselessly by the rarefied hydrogen through which it had to batter its way, it pulled our seas up into a kind of girdle round the equator. Because of the approach of so large a piece of ice, the Earth's temperature dropped disastrously, and terrified humanity had to retreat to the areas which were climatically the least unpleasant. Still the ice-Moon swung inward, until it dominated the entire sky and gave rise to folk-legends of dragons and evil spirits. Finally there came the great cataclysm. The ice-Moon was broken up by the powerful gravity of the Earth, and literally disintegrated. Ice rained down upon the surface of our world, followed by rocks and all manner of miscellaneous débris. The Earth, which had been twisted out of shape, reverted to its original spherical form with devastating suddenness. Earthquakes and violent eruptions followed; the equatorial waters flowed back to higher latitudes, and, needless to say, produced the Biblical Flood.

Obviously we are coming back in spirit, at least, to Dr. Velikovsky, and I am not sure which I prefer: the Velikovsky comet-Venus, or the Hörbiger/Bellamy ice-Moon. I invite you to make your own choice!

According to WEL, the turbulent period following the disruption was succeeded by a long spell of blissful calm. Then, between 13,000 and 14,000 years ago, our present Moon was captured, and there were more earthquakes and volcanic eruptions, one of which submerged the island of Atlantis. (Did you think I had forgotten Atlantis? Not so. I will come to it later.) Again we of the twentieth century are in a period of calm, but sooner or later we must again face tremendous disturbances as our Moon swings inward and is disrupted in its turn.

Such was WEL. Herr Hörbiger was entirely convinced of its truth, and I gather, from people who met him, that it was a little difficult to argue with him. For instance, he stated baldly that the temperature-measurements of the Moon's surface, showing values in excess of 200 degrees Fahrenheit, were deliberately faked. So were the photographs showing that the Milky Way is made up of stars rather than ice-blocks illuminated by the Sun (though, to be honest, it is not easy to tell the nature of a starlike point merely by studying a picture of it; this was a point which Hörbiger did not overlook).

Meanwhile, events in Germany were developing along the lines which led to the emergence of Herr Adolf Hitler and his equally unsavoury satellites. Unquestionably there was a mystical streak in the Nazi hierarchy; astrologers, for instance, were taken very seriously indeed. So was Hörbiger's WEL. Books, periodicals and pamphlets were published in its support; enthusiastic meetings were held, at which orthodox science was reviled and cosmical ice was lustily applauded; and at one time the German Government even had to issue a statement that it was possible to be a good Nazi without also believing in WEL!

Philipp Fauth, who already had a world reputation as a student of the Moon, had already become a Hörbiger devotee, though admittedly he did believe that the Moon had an icy coating rather than being composed of ice all the way through its globe. Astronomers in general tended to regard Fauth's WEL ideas as being out-of-character quirks, and it is certainly true that he spent many years in careful charting of the Moon, though by modern standards his maps are very inaccurate. Fauth regarded the lunar mountains and craters as glacial blocks, produced by the cosmic ice-rain from the depths of space. He did not, of course, suggest that there might be Esquimaux and polar bears living there; but to him the Moon was a very chilly place indeed.

Fauth died during the war. Rather surprisingly, WEL did not die with him. It went on flourishing afterwards, inside and outside Germany, and after the Nazi collapse the Hörbiger Institute re-emerged, to all appearances unharmed. The old-style WEL was modified, but the essential basis was still there as recently as the middle of the 1950s. One of the Institute's pamphlets, published in 1953, claimed that "the final proof of the whole cosmic ice theory will be obtained when the first landings on the ice-coated surface of the Moon take place".

Well—as we all know, the landings have been achieved. When Astronauts Neil Armstrong and Edwin Aldrin stepped out on to the Sea of Tranquillity, on July 21 1969, they had no need of skates. Instead of a smooth sea of ice, blown out from the Sun and deposited on the Moon, their feet met rather crunchy volcanic rock material. Immediately, it became essential to give up the dream of future winter sports on the slopes of the Lunar Apennines. There are no ski-trails on the Moon.

Present-day members of the Hörbiger Institute have accepted this fact, and have come to terms with it. Even before the flight of Apollo 11, the ice concept had been largely given up. In a Hörbiger pamphlet published in 1966, by Mr. Egerton Sykes—a noted researcher in many fields, including cosmology and Atlantis lore—the main emphasis is on Hörbiger's theories of Moon capture, which remain valid. There are also some important speculations as to the nature of the asteroids, those small, barren planets which keep in the main to the region of the Solar System between the orbits of Mars and Jupiter. It is thought that in the remote past, a large planet, Phæton, used to exist there. It was disrupted by a wandering body named Maldek or Mallona, which ran off course, bounced off Pluto and Neptune, knocked Jupiter violently enough to produce the gaping gash we now call the Red Spot, and finally smashed against Phæton, breaking both itself and its victim into a shower of asteroids.

In a revised time-scale, Mr. Sykes estimates that the penultimate Moon crash-landed on Earth about B.C. 25,000, and that we picked up our present Moon in B.C. 10,000; in between these two dates came the flowering of the brilliant civilization of Atlantis. It is still considered that the Ice Ages which have affected our world periodically throughout geological time are linked with the break-up of the various moons; each time such a catastrophe occurs, there is a long spell during which the atmosphere is loaded with débris, so that the sunshine is cut off and the surface temperature drops. This association is, perhaps, the closest remaining link with the original Hörbiger doctrine of universal ice.

I have a feeling that Herr Hörbiger himself would have been less ready to adapt his theories. Probably he would have questioned the authenticity of the whole Apollo programme, and put it down to vile deception on the part of orthodox scientists in their continued attack on the fundamental truth of WEL. But Herr Hörbiger is no longer here to inspire the world with his drive and insight, and we shall never know how he would have reacted. At least WEL has its place in history, and deserves to be remembered even though we no longer contemplate the prospect of steering our future star-ships through the gigantic ice-blocks of the Milky Way.

8

DOWN WITH DARWIN!

Do you believe that men are descended from monkeys? I don't. Neither, for that matter, did Charles Darwin. Many people have completely wrong ideas about what Darwin actually said; but of course he did maintain that you, me, Aunt Agnes and the chimpanzee in the zoo have common ancestors—if you go back far enough. This is a grossly over-simplified version of the principle of evolution.

Most people believe that Darwin was at least on the right track. Others, however, are vehement in their disapproval of the whole idea, and say, uncompromisingly, that "evolution is bunk". Officialdom, both past and present, has been known to adopt this attitude. There was, for instance, the celebrated Scopes trial of 1925. At its height, it was headline news everywhere. I was not interested at the time (perhaps because I was at the early age of two), but it must have been fascinating. Briefly, the State of Tennessee had passed an Act declaring that it was "unlawful for any teacher to teach any theory that denies the

73

story of the divine creation of Man as taught in the Bible, and to teach instead that Man has descended from a lower order of animals". This stern law was deliberately broken by a teacher named Thomas Scopes, who blandly invited the authorities to prosecute him. During the subsequent Court proceedings, the State case was led by one William Jennings Bryan, and the defence was entrusted to Clarence Darrow, probably the most skilful criminal lawyer in American history. Mr. Scopes did not deny his guilt, but when William Jennings was questioned about his reasons for doubting the truth of evolution theory there were legal fireworks in plenty. Not to put too fine a point on it, Clarence Darrow dismembered Mr. Bryan, chewed him up, and then put his remains through a mincer. At the end of the case Thomas Scopes was fined a hundred dollars (a penalty which was later remitted), but nobody could doubt who had been the real victor.

Curiously, the Act was not repealed until May 1967, so that anyone teaching Darwin's theory in a school or college in Tennessee could still have been prosecuted up to a very few years ago. However, I have no doubt that an acquittal would have been secured even without the skill of a Clarence Darrow. Remember, Darwin never suggested that a man is descended from a monkey. As I have pointed out, all he said was that men, apes and monkeys have common ancestors—which is by no means the same thing.

Of course, supporters of the theory of evolution put their faith in fossils, the remains of long-dead creatures found in rocks. Nobody can doubt that fossils are extremely ancient, and go back for millions of years. This puts paid to the theory proposed by James Ussher, a seventeenth-century Archbishop of Armagh, who was convinced that the Earth had been created at ten o'clock in the morning of October 26, B.C. 4004. The good Archbishop arrived at this figure by adding up the ages of the patriarchs, and making some other calculations of the same kind, which pleased the theologians but did not satisfy the fossil-hunters. However, his view was officially accepted for many decades after his death in 1656. Seventy years later a German scientist, Professor Johann Beringer of Würzburg, published a book in which he claimed that fossils were of divine origin. Actually, the fossils concerned had been faked and planted by the students in his University class—a fact which he finally realized when he unearthed a fossil which had his own name inscribed on it. (According to legend, he then did his best to buy up all surviving copies of his book—without much success.)

Darwin's famous *Origin of Species* came out in 1859. In it, he proposed the now-celebrated idea of natural selection, which can equally well be called "the survival of the fittest". Creatures which are unsuited to their environment will not persist; they will die out, leaving the field to others better able to cope with the situation. For instance, suppose that you have two races of giraffes, one with long necks and the other with short necks? The tastiest leaves are on the tops of the trees. Long-necked giraffes will be able to reach them, and can nibble away to their hearts' content. Short-necked giraffes will be unable to share in the good food; and since they cannot live on morsels to be found lower down, they will die out. Eventually, only the long-necks will be left. Nothing could be more clear-cut.

On the basis of evolution, the story of Man began a very long time ago. In the so-called Age of Reptiles, between about 200 and 70 million years in the past, the world was dominated by vast creatures beside which our modern frogs and lizards would seem very puny. Consider the Tyrannosaurus, which was at least fifty feet long, and carried its head some twenty feet off the ground. Its mouth was even larger than that of a modern politician, and was equipped with a row of six-inch teeth sharper than swords. It can hardly have been a friendly creature, but no man has even seen one, because all the huge dinosaurs died out

70 million years ago, and the story of *Homo sapiens* does not go back anything like that far. We can safely discount any picture of Mr. and Mrs. B.C. locking up their cave at sunset and putting the iguanodon out for the night.

Natural selection tells us that the dinosaurs bowed out because the general environment changed, and conditions became unsuitable for them. As they were not distinguished by their cleverness—their brains were smaller than those of a modern kitten—they simply faded away. Meanwhile, mammals had begun to develop, and in the tree-tops there pranced around some rather special mammals known as primates. It is from these, say the Darwinians, that we are descended. So are apes and monkeys; but they come down according to a different line, so to speak, and will not evolve any further. There is no reason to suppose that an ape will ever acquire human traits (at least, one hopes not, for the sake of the apes), and neither is it true to say that any modern ape or monkey is an ancestor of ours.

Even before Darwin, of course, there had been suggestions of a real relationship between man and monkey. The Scottish anthropologist Lord James Monboddo, who died in 1799, believed that orang-outans belonged to the human species; he also maintained that all human children were born with tails, duly removed immediately after birth by conscientious midwives. But it was only with the rise of Darwinism that arguments began to rage far and wide. Religious objectors were much to the fore, and there was the famous debate between Huxley, one of Darwin's leading supporters, and Wilberforce, the Bishop of Oxford. It ended in the rout of the Bishop. Huxley even said that if he had to choose an ancestor, he would infinitely prefer an ape to Mr. Wilberforce.

There are still a great many religious objectors to any kind of evolution. The Roman Catholic Church, for example, is frankly glossing over the entire problem; it doesn't want to accept Darwinism, but cannot really see a way out. However, let us turn to the purely scientific arguments of those who call "Down with Darwin!" Some years ago I discovered that there is a fully-fledged Evolution Protest Movement, with its headquarters in Hayling Island in Hampshire, England. Its leading light is Mr. A. G. Tilney, who was for many years Senior Languages Master at an important grammar school for boys in Portsmouth. Since retiring from teaching, in 1955, Mr. Tilney has been Secretary of the Movement, and has spared no effort to propagate its ideas; he has visited most of the countries of the world, speaking and broadcasting; and he has many followers.

I have had several long talks with Mr. Tilney, and much of what follows is drawn from our conversations, the latest of which was only a few months ago. It is quite in order to describe him as an Independent Thinker, since his opinions are so much at variance with those of orthodox science. Mr Tilney agrees with this, and welcomes it. He does not disregard the purely religious view—far from it —but let us deal here with pure science.

The first argument is that evolution goes against the second law of thermodynamics. Energy is transferred into non-reversible heat energy, and the energy loss is referred to as entropy. Now, entropy must always increase. As Mr. Tilney has said, "All things left to themelves tend to go bad, sour, mouldy, rotten and disorderly." In many walks of life this is true; it is particularly true of our present-day schools and universities, where entropy had been accelerating ever since the psychiatrists assumed control of state education after the last war. Also, look what happens if you leave a piece of cheese in the larder and forget about it. But the theory of evolution teaches that more complex and organized creatures are produced from more primitive ones; our horses from the tiny eohippus, our elephants from small Tertiary mammals, and, of course, men from apes. (The Tertiary era covers the geological period from 70 million years ago, when the dinosaurs made their exit, up to about one million years ago.) All this, says Mr. Tilney, is basically unscientific.

We must also consider the Ice Ages, in rather a different light from the way in which Herr Hörbiger did. Undoubtedly there have been periods when the temperature of the world has been much lower than it is now, and it is generally thought that the cause has been due to slight variations in the light and heat output of the Sun, though different authorities have different ideas about it. The most recent series of Ice Ages began around 750,000 years ago, and ended a mere 10,000 years ago—that is to say, B.C. 8000, when the ice-cap retreated for the last time, and the rise in sea-level flooded the area now separating Britain from the mainland of Europe, producing the North Sea and the Channel. Before then, a man could have walked from Dover to Calais, or, for that matter, from Grimsby to Oslo. After B.C. 8000 Europe was cut off from England, though its inhabitants were probably quite unaware of their misfortune.

Mr. Tilney and his colleagues maintain that no man could possibly have survived the Ice Age cold, so that *Homo sapiens* cannot go back as far as B.C. 8000. Instead, men were suddenly created about 6,000 years ago—that is to say, around B.C. 4000, which brings us back to Archbishop Ussher even if for rather different reasons. These first men had no ancestors. It may even be said that they popped up out of a trap, as it were. Subsequently there came the Flood, which was not due merely to an exceptionally wet season, but to a sudden earth-tilt which brought the waters bursting forth from below the crust. We can also show that at that time the different kinds of animals were not so numerous as they are today. There would have been no room for them inside the Ark!

I admit that I can see a difficulty here. If there were fewer kinds of animals at Flood-time, then the new varieties must have been created since, in view of the Movement's opinion that no creature can ever evolve into another kind of creature. Each species is created by a special act. If, therefore, you happen to go for a walk one day, and come across an animal with a dog-like body, trotters like a pig, and large, flapping ears, you will know that there has been a new creation.

Before going any further, let us deal with those darned giraffes. Far from proving Darwinism, they argue against it. As Mr. Tilney points out, the female giraffe is about two feet shorter than the male, so that if natural selection had operated it would have been the females which would have died out. This would have led to the extinction of the whole species, for rather obvious reasons. It is also stressed that giraffes, as a class, are very fond of acacia trees, which are within convenient reach; and the giraffe prevents the acacia trees from becoming too numerous. As Mr. Tilney says, it is "an example of Nature's pruning". In my old home at East Grinstead, England, we had a couple of acacias in the garden, and I admit that I did not notice any tendency to wild spreading; but if they had begun to multiply alarmingly, no doubt I could have borrowed a giraffe from somewhere or other.

Having disposed of Darwin's natural selection, we must also deal with the older theory due to the French scientist Jean Lamarck, according to which animals were able to adapt themselves, over many generations, to cope with their needs. On Lamarck's principles, giraffes would have found it necessary to keep on stretching upward at dinnertime, so that their children would have acquired necks which steadily became longer and longer. (I apologize for this giraffe surfeit, but it really does give an excellent illustration of what is meant.) But this is also illogical, and Mr. Tilney cites the case of the spider, which has a complicated mechanism for spinning webs. This web-spinning device would be quite useless until fully developed, and in the intervening period—when the spider was gradually developing its ability, and becoming thoroughly web-minded—the relevant organs would have been so much in the creature's way that the results would certainly have been fatal. Therefore, the spider must have been created exactly as it is now, with its web-spinning organs in full working order.

Moreover, no evolution is taking place today. Orthodox scientists maintain that the lack of obvious change is because the whole process is so slow—except in a few cases, as with particular kinds of moths. This argument is completely discounted by the Evolution Protest Movement, and it is also said that no artificial species of animal or plant has ever been much of a success. Mr. Tilney gives the case of a Russian geneticist, Karpenchenko, who in 1924 crossed a radish with a cabbage, producing a plant which he called Rephanobrassica (though I would prefer to name it either a caddish or a rabbage). Unfortunately, the result combined the worst characteristics of both its parents. It was also sterile, so that in any future war we are in no danger of having to eat rabbage pie.

Having cleared the air, we must turn to the most important aspect of all: the record of the rocks. Orthodoxy claims that there is a complete geological story, beginning over 500 million years ago with the first recognizable fossils, and coming right through to the present day: simple, single-celled creatures, fishes, amphibians, reptiles, primitive mammals, more complicated mammals, primates, ape-men, true men. Seen in museums, these classified fossils look very imposing, but the E.P.M. is far from impressed. Mr. Tilney is quite definite that there can be no "half and half" creatures, and that the record is so incomplete that it is no more than guesswork. "There is an unbridgeable gap between the age of fossils and the living world of today." It is maintained that evolution-minded scientists can take a tiny fragment of something or other and then build up an imposing picture which may be nowhere near the truth. For instance, it is correct that one famous painting, showing two ape-men cavorting round their camp fire, was worked out entirely from the evidence of a single tooth found in the State of Nebraska, which subsequently turned out to be that of an extinct pig. There is also the famous—or infamous—Piltdown Man.

Piltdown Man first appeared upon the scene in 1912, when Charles Dawson, an amateur archæologist and palæontologist, produced a skull and jawbone which he said he had found in a gravel-pit near the village of Piltdown, in Sussex, England. The interesting feature was that the skull was definitely human, the jaw just as definitely ape-like. Here, it seemed, was the true Missing Link; and Dawson's report to the London Geological Society, presented together with support from a well-known expert named Smith Woodward, caused a scientific upheaval. The age of Piltdown Man was estimated at half a million years, which would put him back into the middle of the Ice Age. In 1915 Dawson produced some more bits of skull, together with a tooth. I have a textbook, written in 1927, which states that "we have here a type very close to modern man, though the ape-like character of the jaw remains a puzzle".

Dawson died a few months after his last discovery, and for the next few decades Piltdown Man remained an enigma. In November 1953 he was back in the news with a vengeance. Geologists at the Natural History Museum, London, announced that they had made a careful re-examination of the remains, and had used new, improved methods of dating which had been quite unknown in Dawson's time. The results were only too painfully conclusive. Piltdown Man was a fraud. The skull was human, and of moderate geological age; the jaw was that of a modern ape which had probably been alive and kicking not so long before it had turned up in the gravel-pit. The various tools and animal bones found nearby had been carefully stained and treated to make them look much older than they really were, and the teeth which Dawson had produced had been expertly treated by someone with a sound knowledge of dentistry. Older residents recalled that Charles Dawson had had dental apparatus fitted up in the attic of his house. . . .

Piltdown Man, or Eoanthropus (Dawn-Man) was dethroned. The site of the "discovery" was removed from the official list of Ancient Monuments, and today the only recognition in the village is the local pub, which is still known as *The Piltdown Man.* Scientists, like Queen Victoria on a previous occasion, were not amused. If Mr. Dawson had still been alive, he would have been the subject of some rather caustic remarks. But as the Evolution Protest Movement says, what can happen once can happen again—or even a hundred or a thousand times. To put too much faith in fragmental remains is, surely, extremely dangerous.

Not, of course, that Piltdown Man was the only specimen of supposed primitive humanoids. There is Pekin Man, also belonging (it is officially claimed) to the Ice Age. He was unearthed in 1927—or, rather, his skull was—and he proved to have a flat head and large teeth. Whether he was an expert swimmer in the Chairman Mao class is not known. I need not catalogue all the other fossil men; according to the E.P.M. they are, without exception, either fakes or errors. Neither can we have any time for the still more primitive creature, Australopithecus, of at least a million years ago. This would take us back before the Flood, when no men could exist anywhere.

To the E.P.M., the "dead hand of Darwin" is the worst of our modern evils. It is even linked with Marxist Communism and with the inflammatory music of Wagner. Protests have been sent regularly to the British Museum, London, for its publication of evolution-theory propaganda; to television companies for various programmes; and to many other bodies. For some odd reason, these protests have fallen upon deaf ears. Nonetheless, Mr. Tilney and his colleagues continue with their work, with the eventual aim of making sure that "biology be taught in universities, schools and on the radio as a science, instead of in the form of evolutionary propaganda as hitherto".

There, for the moment at least, the matter rests. Quite apart from the religious overtones, the Movement puts forward scientific arguments which, they say, cannot be refuted. The idea that a monkey, an ape or any other kind of animal may be a distant relative of ours is, to them, as illogical as it is repugnant. There is an absolute gulf between Man and any other kind of creature, and this gulf will never be bridged. True, the E.P.M. spokesmen do not go as far as a Mrs. Hackett, who initiated a remarkable correspondence in 1967 in the columns of the leading Northern Ireland daily paper, the *Belfast News-Letter*; she announced that on Christian principles she always fastened up the swing in her budgerigar's cage on Saturday evenings, so that it would be unable to disport itself on a Sunday. All the same, the members of the Evolution Protest Movement are among the most dedicated, the most sincere and the most energetic of all modern Independent Thinkers.

9

UP ATLANTIS!

It is never easy to draw a line between history and myth, and King Arthur is an obvious borderline case. During the period between the undignified departure from Britain of the Romans and the rise of the Seven Kingdoms, Arthur—or someone to whom we now refer as Arthur—probably existed; he may even have held up the invaders from Europe for some time. This is history. Alas, the Round Table and the noble band of do-gooders such as Sir Lancelot, Sir Bedevere and the rest belong to myth. I am very fond of Tintagel, Glastonbury and other Arthurian sites; but I am very dubious as to whether any Round Tables existed there before the advent of modern tea-shops.

However, even the fascinating Arthur pales before the magnificent civilization of Atlantis, which—so it is said—flourished many centuries ago, and was finally submerged, sinking dismally below the surface of the sea in a remarkably short time. The date of its destruction is given variously by various authorities, but B.C. 13,000 is probably not far from the mean estimate. This means that it must be classed as prehistoric, and we are back again in the final part of the Ice Age, which, remember, ended about B.C. 8000.

I do not propose to give all the background to the Atlantis story, partly because I am not setting out to describe either myths or pre-history, and in any case it would mean writing a book the size of an encyclopædia. I have no idea how many volumes, pamphlets, articles and columns have been written about Atlantis; the total is certainly many thousands. I mention them here because there is still an Atlantean Society whose members certainly rank as Independent Thinkers, and who believe that the vanished civilization was the greatest that Earth has ever known. There is also a strong link with flying saucery, which will emerge in due course.

Let us clear the air first, and dispose of two other legendary civilizations; those of Mu and Lemuria. Sometimes they are separated; other researchers consider them identical—as did their leading publicist, the late Colonel James Churchward of the Bengal Lancers. Apparently Mu was a huge continent, located somewhere in the area now occupied by the South Seas. Its inhabitants—perhaps it is better to call them Lemurians rather than Mueys—were remarkably skilled in technology; for example, they had mastered anti-gravity devices, so that they could fly gaily in the manner of birds. Churchward even suggested that this was how Jesus managed to walk on the water, a feat which, so far as we know, has not been performed since (though I have little doubt that at least one famous modern British Prime Minister has tried it). Then, unforunately, there was a gigantic explosion, probably triggered off by the spontaneous ignition of gases trapped below the Earth's crust. Mu sank out of sight, taking all its teeming millions with it—apart from a few who somehow or other managed to get away, to end up later in Atlantis.

The same sort of idea was supported by Madame Blavatsky, founder of the cult of Theosophy, and her successor, Mrs. Annie Besant. Both these ladies considered that Mu-Lemuria was the first great civilization, in which men were created. Originally they were part-physical and part-astral, but later they became only too human, and had acquired all the familiar traits before the big bang occurred and the survivors escaped to Atlantis. Naturally, there is much more than this to the study of Theosophy, and I have read many publications of the modern Theosophical Society—which I do not propose to consider here for the excellent reason that I still have not the faintest idea of what they are all about.

Now let us leave Mu and come to Atlantis. Our source here is no less a person than Plato, the great Greek philosopher who lived 400 years before Christ. He wrote voluminously, but for the moment we are concerned with only one book—and an unfinished one at that. It is called *Kritias* (or *Critias*; one has to be phonetic, and there is some uncertainty, owing to the prevalent Greek habit of using strange letters such as alpha, beta and gamma). *Kritias* is in the form of a conversation held between Solon, the eminent Greek lawyer, and Kritias himself. The conversation is set in Egypt, around B.C. 590. So far we are on firm ground; there is excellent evidence that Solon did go to Egypt around that time, and of his historical existence there is no reasonable doubt at all.

The Kritias of the book is the grandfather of the narrator, and apparently the story of Atlantis was old even in his day. He describes the island as being larger than Libya and Asia put together (by this he presumably means Asia Minor); he says that it was the site of a "great and wonderful empire" whose power extended right out as far as Egypt and Italy. The only people to stand against Atlantis were the Athenians, who had rather the better of matters from a military point of view. Eventually "there occurred violent earthquakes and floods"; in a single day and night Atlantis vanished forever below the water, together with the main Athenian army. Kritias added that the sea in those areas became impassable, because of the shoals of mud left behind by Atlantis.

Let us concede that islands—even large ones—do appear and disappear sometimes. What we have to decide is whether Plato's account is fact, part-fiction or pure fiction. In his story, Atlantis was originally created by special order of the gods, and was given plenty of what we now call natural resources. In particular there was a metal or alloy named orichalcum, of which we know only that it had a reddish tinge. The design of the island-empire and its capital city is described in detail, as are some of the customs and the public amenities; for instance there were separate swimming baths for women, designed presumably by some Atlantean moralist. For a time all went well. The original Atlanteans had been partly Olympian, but as generation succeeded generation the human strain began to dominate the celestial influence. The Atlanteans started to behave like men instead of gods. This did not please Zeus (or Jupiter, if you like; but we are talking about Greek gods, not Roman ones). The end product was an Olympian-caused cataclysm which solved the problem once and for all by simply removing the entire island from the face of the earth. It was admittedly a drastic remedy, but in general the Olympians were not noted for their kindliness and tolerance.

Plato's manuscript is unfinished, and breaks off in the middle of a sentence, rather in the manner of a modern television play. It remains our only real source material for the existence of Atlantis, but it was seized upon by many later writers, and in particular by Ignatius Donnelly, one-time U.S. Congressman, State Senator, and Presidential candidate. It was Donnelly's book *Atlantis: the Antediluvian World*, published in 1882, which made the story sound really credible. He assembled a mass of material in an attempt to show that there are similarities between the geology, archæology, architecture and folklore of places ranging from Egypt to Mexico—and that these prove that all came from one source, which can only have been Atlantis. He added that the kings of Atlantis were identical with the pagan gods of later civilizations.

Donnelly was nothing if not persuasive. One convert was no less a person than W. E. Gladstone, then Prime Minister of Britain. Soon after the publication of the book, Gladstone asked the Treasury for funds to send a ship to the suggested site, in the Atlantic, and try to trace the outline of the sunken isle. In those days, however, the Treasury was much more responsible than it is today, and was much less inclined to waste public money upon futile projects. Gladstone's request was turned down, and nothing more was heard of it.

Donnelly was not concerned solely with Atlantis. He believed that in the past the Earth had been devastated by a huge comet, after which there had followed the Ice Age; Atlantis arose later, and there is only a loose link with Dr. Velikovsky's comet, because the time-scale is different. Finally, Donnelly turned his attention to Shakespeare, proving that the plays contained a cipher showing that the author was really Francis Bacon. My own views on this are clear-cut. It is transparently obvious to any scholar that William Shakespeare's plays were written not by William Shakespeare, but by another author of the same name. However, this is getting rather far from Atlantis, and to bring us back to the point I must return by way of Hans Bellamy, whom we remember as a disciple of Hörbiger's cosmic ice theory or WEL.

(It is amazing how the Independent Thinkers are intertwined. Donnelly to Bellamy, Bellamy to Hörbiger, Hörbiger to Velikovsky, etc. . . . only in a few cases, such as that of Mr. Bradbury, do we go off at a complete tangent.)

Bellamy looked around for a possible scientific cause of the destruction of Atlantis, and he found it in WEL. As we have seen, according to WEL the Moon used to be independent, and was captured by the Earth at just about the time of Atlantis. The new satellite's gravitational pull forced enormous masses of water into the tropics, and kept them there for some time, thereby submerging low-lying islands and coastal zones. Atlantis was one of the casualties, and was submerged together with its temples, its palaces, its inhabitants and its highly respectable and segregated public baths.

So far as topography goes, we must come back to Plato, who described the island as oblong, with a steep coastline. The capital was architecturally interesting, with various concentric rings of land and harbour; in the middle of the inner citadel was a temple dedicated to Poseidon, the sea-god who had founded Atlantis. (You may know him better as Neptune.) In the temple were ornaments of silver, gold, ivory and orichalcum. Laws in Atlantis were applied with firmness and justice; there were rich and varied fauna and flora, including elephants; and the climate was mild and pleasant. It all sounds positively Utopian.

But where *was* Atlantis, if anywhere?

The obvious location is in the Atlantic, and Donnelly thought that the Azores are the tops of sunken Atlantean mountains. But many other sites have been suggested. Among them are Morocco, Nigeria, Heligoland and the South Seas. Atlantis seems to have been everywhere at some time or other; when astronauts reach Mars, I have not the slightest doubt that they will find traces of Atlantis there too. But the most favoured theory among modern scientists of the orthodox school is that the story relates to the Mediterranean island of Crete, once ruled by the legendary King Minos.

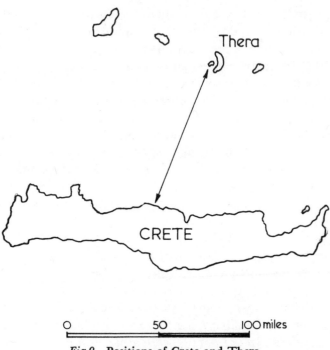

Fig 9 Positions of Crete and Thera

Nobody doubts that early Cretan civilization was highly developed, whether or not we accept the existence of Minos himself. About eighty miles away from the island lies Thera, or Santorin, which is nothing more nor less than a volcano of the type which occasionally goes off with a devastating bang. According to unimpeachable geological evidence, this happened in or around B.C. 1500. The effects on Crete were cataclysmic. The island was not sunk, but its coastal regions were swamped, and the death-roll must have been tremendous. The so-called Minoan civilization never recovered, and eventually disappeared altogether.

This would have happened some nine hundred years before Solon's visit to Egypt as described by Plato at a still

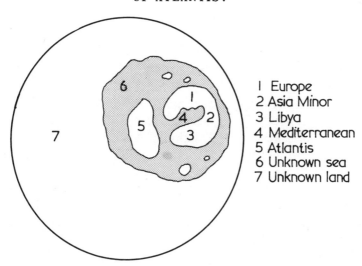

1 Europe
2 Asia Minor
3 Libya
4 Mediterranean
5 Atlantis
6 Unknown sea
7 Unknown land

Fig 10 Ancient map of Atlantis

later period, and so the details of what happened would have become very blurred. It is perfectly logical, then, to link the Cretan tragedy with the legend of Atlantis. In modern times we have had experience of what outbreaks of this sort can be like. In 1883 Krakatoa, in the East Indies, behaved in similar manner; the original island was destroyed, thickly-populated coasts of other islands were flooded wholesale, and over 30,000 people lost their lives. The dust produced from the outbreak hung in the upper atmosphere for three years, and the tidal waves set up were measured as far afield as the English Channel. Krakatoa is the same species of volcano as Thera, and no more destructive even when it really puts its mind to the job.

Historically, this is about as far as we can go. We know much more about the ocean-bed nowadays than Gladstone did, and it needs considerable wishful thinking to see the outline of Atlantis anywhere in the Azores area; the other sites are equally unpromising, and after so many centuries it is hardly likely that there is any undiscovered evidence about the liquidation of Minoan Crete by the vagaries of Thera. Therefore, modern Atlantean enthusiasts have to rely upon less tangible forms of evidence, many of which come in the form of messages from the Great Beyond. One, a book written in 1957 by H. C. Randall-Stephens, is in the form of messages sent through to him by an initiate of Ancient Egypt. There are also various societies, one of which holds that the destruction of Atlantis—caused by the sudden tilt of the Earth's axis afer the capture of our present Moon, à la Hörbiger—will be paralleled by events in the future. In an issue of *The Atlantean*, unfortunately undated but probably published about 1960, it is said that the main threat comes from the planet Uranus, which has already started to break up, and whose fragments can sometimes be seen in the form of comets. In the foreseeable future, Uranus will slip from its normal orbit and "run amok" through space, burning as it goes. As it by-passes the Earth, it will cause a violent switch-over in our axis; there will be a period of darkness, as the rays of the Sun are blocked out; black rain will fall from the skies; the poles and the equator will change places; old lands will vanish, and new ones will appear. England, I am glad to say, will benefit from the change, and after the chaos it will be larger than of yore, with a climate similar to that of present-day South Africa. It is also good to know that the enlightened inhabitants of the planet Venus will contact us and help us to rebuild our civilization. So will the denizens of Jupiter and Pluto, though these may be less efficient. (Another issue of *The Atlantean* gives a vivid description of life on modern Pluto, at a distance of over three thousand million miles from the Sun. Because of the prevailing gloom, artificial lighting there is very highly developed, and we also find that the inhabitants have perfected an elaborate system of underground railways.)

I am not straying from the point once more, because it is also claimed that in this general period of upheaval

Atlantis will again emerge from below the waves. Its evolution, we are told, was never completed, and its destruction was premature. (The Society dates the submersion as B.C. 5000, but this may well be a misprint for B.C. 15,000; I have been unable to find out for certain.) So Atlantis will be re-born, as we are assured by the sender of several messages, a gentleman from the Beyond with the impressive name of Helio-Arcanophus.

Other researchers believe that the trouble came not from Zeus, or even from Thera, but from the Atlanteans themselves. As time went by, and they fell further and further from their initial Golden Age, they were ill-advised enough to start tampering with the structure of the atom. This was both regrettable and unnecessary, since they had already solved the problem of anti-gravity; but all the same, they did it. Eventually something went wrong. It may have been an over-hasty preparation of a weapon of war. It may have been due to some Atlantean scientist, in his laboratory, pushing the wrong button. In any case, the result was an explosion beside which even Krakatoa would look like a squib. Atlantis (or Mu-Lemuria) disintegrated; parts of it were hurled clear of the Earth altogether, and became fragments circling the Sun in their own independent orbits, still carrying their surprised inhabitants. We know these fragments today as the asteroids. Other pieces of the disrupted isle were hurled as far as Venus, and the present-day Venusian civilization is of Atlantean origin; possibly the same can be said about Mars. Meanwhile, the situation on Earth was very far from being under control, and not only Atlantis but also the rest of the world was devastated by the big bang. Of course, this was the Biblical Flood, and it was only after a very wet period that things began to return to normal—minus Atlantis.

Orthodox science will question many of these concepts. Uranus, for instance, is believed to be a giant world in a perfectly stable orbit, never approaching us to within a distance of much less than 1,500 million miles; it is debatable whether any underground railway network would be practicable on Pluto, where the temperature is so low that air of the kind we breathe would turn into liquid; and a super-bang able to send parts to Atlantis far into the empyrean would have rather unhappy effects on anyone who happened to be there at the time. The shock would be decidedly too great to be passed off with a light laugh and a fresh cup of coffee. However, the modern Atlanteans base their views partly upon the Ancient Teachings and partly upon messages from higher spheres, so that strictly scientific arguments can safely be discounted.

There is another loophole also. If the Venusian civilization came from Atlantis, it could well be that most of the top Atlanteans went there voluntarily as soon as they knew that their homeland was about to be swamped. And this leads us on to the fabulous vimanas, which may be linked with flying saucers.

Most of my information here comes from my old friend Desmond Leslie, co-author of the famous book *Flying Saucers Have Landed*. It is held that there is a close association between gravity and magnetism, and that the Atlanteans, as well as the Lemurians, had learned how to raise their vibrations above those of the Earth's cold magnetic force. (Please do not ask me what this means, or how it could be done!) Once the magnetic force is nullified, the body concerned has no weight, so that gravity no longer applies. This is also the principle behind levitation and also, no doubt, the Indian Rope Trick.

As soon as the secret of anti-gravity had been discovered, making flying machines became child's play, and those of the Atlanteans were highly sophisticated. They were called vimanas, and were either round or boat-shaped, with twin rows of nozzles through which their power was sent out. They glowed in the dark, and were made either of metal (orichalcum?) or else of wood cemented together by a long-lost chemical process which made them remarkably tough. Inside they were very comfortable, with every modern convenience apart possibly from a cocktail bar.

When the crisis became imminent, the enlightened rulers of Atlantis held a conference, and decided to get out while the going was good. Accordingly they made their adieux, boarded suitable vimanas, and set off for Venus. The Saucers which visit us today, maintains Desmond Leslie, are carbon copies of the Atlantean vimanas. They may even be the original ships, which were presumably made with great care and would not become obsolete in a mere fifteen thousand years or so.

Can we ever prove the existence, or non-existence, of Ancient Atlantis? Not unless we can establish contact with beings who are much more advanced than ourselves, and this may not be entirely easy. We must hope that our Atlantean enthusiasts will find a way round the problem. But having introduced the subject of flying saucers, perhaps we had better discuss them more fully—because it is in the realm of Flying Saucery that we find the best-known of all Independent Thinkers today.

10

CROCKERY FROM
THE VOID

"As I studied the magnified surface of the Moon upon the screen...a small animal ran across the area I was observing. I could see that it was four-legged and furry, but its speed prevented me from identifying it."

A report from one of the Apollo astronauts? Alas, no. Up to the present time no space-expedition has found any trace of life on the Moon. Airless and waterless, the lunar world is not particularly welcoming, and most people (including myself) believe that it has always been sterile. However, not everyone concurs, and one of the dissentients was the late Mr. George Adamski, who wrote the description quoted above. It was published in 1955, in a book called *Inside the Space-Ships*. At the time, Adamski relates, he was having a trip in a flying saucer piloted by an astronaut from the planet Saturn, and they were not very far away from the actual surface of the Moon.

I have chosen George Adamski to introduce the subject of flying saucers chiefly because I knew him (he died, unfortunately, a few years ago). I also knew another of the main characters in the saucer saga, Cedric Allingham. But first we must go back to June 24 1947, because it was then that the story first hit the headlines. The observer concerned, Kenneth Arnold, was flying a private aircraft somewhere over the State of Washington when he saw nine round objects, in obvious formation and moving at high speed, passing within a few miles of him. He said that they were "flat, like pie-pans", and apparently the term "flying saucer" came from this.

Arnold believed that he had seen space-ships from another planet; he also believed that the Authorities (with a capital A) were dedicated to playing down his story and suppressing the great truth. Before long there came other reports, some of them even more sensational. If they were to be credited, then the Earth was under close scrutiny, though by whom or for what purpose remained obscure. Neither was it known whether the Saucers were friendly, hostile or merely inquisitive. Photographs of them were invariably and inevitably blurred, and their whole behaviour could only be classed as erratic. Many people dismissed them as being due to mundane objects such as birds, bats, balloons, aircraft, auroræ, searchlight beams or even the planet Venus. In the early days I remember being telephoned at 3 a.m. by an agitated lady living not far from my home, which was then at East Grinstead in Sussex. She announced that she had seen a flying saucer, moving at great height and pausing to hover over her garden. When I asked how it was moving, she said that it "appeared to be flapping". I had no doubt it was doing just that, but I was totally unable to convince her that she had been watching the flight of an early bird on the look-out for its traditional worm breakfast.

Where there are space-ships, there must presumably, if not necessarily, be astronauts; and it was more or less axiomatic that someone would bring in a report of contact with extra-terrestrial beings. The spark was ignited by George Adamski, who published the famous book *Flying Saucers Have Landed* in 1953. His co-author was Desmond Leslie; but Desmond's section was confined to historical accounts, comments and interpretation, because he had not then seen a Saucer for himself, and his original contact with George had been by telepathy. (In fact Desmond has yet to see a Saucer really close-up, as he is the first to concede.) The really sensational part of the book was the Adamski account of his meeting with beings from Venus, which, he told us, took place at 12.30 p.m. on November 20, 1952.

The place was near Parker in Arizona, actually on the Californian desert. Together with some friends, Adamski saw the Saucer; alone, he located the pilot, who was rather like an Earthman but for the fact that his clothing was rather atypical for a desert expedition. For instance, he was wearing what seemed to be ski trousers. Also he had shoulder-length hair, beautifully waved. This would seem quite run-of-the-mill today, when our universities are cluttered up with unwashed creatures with long hair and whiskers; but it was out of the common in 1952, just as we hope it will again be out of the common in, say, 1982.

Unfortunately the visitor would not talk English, and George had to communicate in semaphore, but he was able to establish that the Saucer had come from Venus. At the end of the interview the Saucer itself came along to join the party, and hovered above the ground. Inside it were several other people. It was in this remarkable craft that the visitor made his final exit, leaving George alone in the desert.

When the book appeared, it became an immediate best-seller. I have no idea how many copies were distributed; it must have been around the million mark. George Adamski became a world figure almost overnight, and remained so until the time of his death more than ten years later. When he came to London, I met him; indeed, we even talked together on a British television programme called Panorama. And let me stress that there is no doubt in my mind that he meant everything he said. He was a true Independent Thinker, and he was quite sure that he had been singled out for special attention from the Saucerers.

Three years later he followed up his book with another, in which he described how he met three Saucer pilots in the lobby of an hotel, and was taken for a ride. It was then that he soared near the Moon, and saw the amazing scenes quoted at the start of this chapter. One of the crew—a Martian, this time—even explained how the Saucers worked. The written description of the Saucer originally given by Adamski, and also his remarkable published photograph of it reminds one of a lampshade with three bulbs underneath (suggestions that it *was* a lampshade with three bulbs underneath I dismiss as unworthy). According to the Martian, these are hollow balls which collect and condense the static electricity sent to them from the magnetic pole. This power, of course, is present everywhere in the universe; one of its natural but concentrated manifestations is known to us as lightning.

Altogether George went for several flips, both in this and in a larger, more ambitious craft built upon the planet Saturn. By this time his hosts had decided to speak English rather than waste time in semaphore, and all sorts of illuminating facts emerged. Eventually, after several such episodes, the interplanetary visitors departed for home, though not before George had been the guest of honour at a farewell banquet held in the Saturnian Saucer.

The last Adamski book, *Flying Saucers Farewell*, appeared in 1961, but on the whole it did not add much to what had been said before. In the meantime, Cedric Allingham had come upon the scene, flitted through the pages of Saucery in the manner of a meteor, and left again as mysteriously as he had arrived.

This time the event took place in Scotland, on the coast between Buckie and Lossiemouth. Mr. Allingham, an amateur scientist, had been on a caravan holiday, watching birds (the feathered variety), when an Adamski-pattern Saucer put in an appearance and a gentleman in a one-piece space-suit stepped out, holding his arm aloft in salute. Like the Californian visitor, he spoke no English; but all the same, he and Cedric managed to cope with the situation very well.

It was soon established that this particular Saucer did not, like Adamski's, come from Venus. It had been built on Mars; and since the air there is much thinner than ours, the visitor had to breathe by means of a tiny tube stuck up each nostril—presumably some form of oxygen apparatus. It was also found that he had been to Venus, and that the Martian canals are true waterways; at that time, remember, Earth-launched probes to Mars were still a novelist's dream. The visitor did not object to being photographed, and the picture, published in Mr. Allingham's subsequent book *Flying Saucer from Mars*, proves that the Martians, like ourselves, have to wear braces to keep up their equivalent of trousers.

The sad thing about this encounter was that there were no witnesses to it except for a lighthouse-keeper named James Duncan, who produced a written statement, but whom nobody ever seems to have tracked down from that day hence. This weakness was freely admitted by Cedric Allingham himself, as he told me when I met him some months later at a private lecture he delivered to a flying saucer group in Tunbridge Wells, England. I never saw him again, and there were reports that he had died abroad —unless, of course, he had made fresh contact with the Martians, and is at this moment basking in the weak sunshine along the banks of some canal many millions of miles away from us.

Cosmical visitors, as described by Adamski and Allingham, would certainly be welcome anywhere; but there were other suggestions of a more sinister kind, and it was even hinted that the Saucers might be hostile, in which case they would presumably have been dispatched Earthward by different races. Some books written in America by a Major Keyhoe claimed that the Authorities (still with a capital A) knew all about the space-ships, and were doing their nefarious best to conceal the truth. During the late 1950s and early 1960s the stories continued to come in, and Saucer Societies sprang up like mushrooms. My favourite report concerned a small Saucer which crash-landed in Mexico, killing the crew; on examination the occupants were found to be little green men. Later there was a correction; the little men were not green at all, but bright red. Finally it transpired that the whole story had been put out by the manager of the only local hotel, who did a roaring trade for several days while journalists and reporters from all parts of the United States swarmed into the area!

It was around this time, too, that a serious split occurred in the ranks of Flying Saucery. The Independent Thinkers divided themselves into two distinct camps. There were the "contacts", following Adamski and Allingham. Then there were the more cautious investigators, who dismissed these contact reports as being due to hallucinations or hoaxes, but who still maintained that the Earth was under surveillance. The very term "Flying Saucer" was tacitly dropped, to be replaced by the much more imposing title of Unidentified Flying Object, or U.F.O.

So far as I was concerned, the next highlight came in July 1963, when some peculiar craters appeared in a potato field at Charlton—not the London Charlton, but a small village near Shaftesbury in Dorset, England. A local farmer made the discovery, and also saw that the crops for a wide area around had been flattened. Reports over the radio and in the Press caused widespread interest, and this was heightened by a statement from an Australian who gave his name as Robert J. Randall, from the rocket proving ground at Woomera. Dr. Randall maintained that the crater had been produced by the blast-off of a Saucer from the planet Uranus. It was independently suggested that there might be a bomb in the crater, and an Army disposal squad was called in.

When the whole affair started to look really interesting, I happened to be in a television studio. We decided that whatever had happened, we ought to be "in on it". At the dead of night we drove to Charlton, and arrived in the early hours to find all sorts of people hopping about like agitated sparrows. The bomb disposal squad was at work, but had unearthed nothing but a piece of metal which might have been anything. As I was once involved with dismembering bombs (as a passing phase during the war, while I was with the R.A.F.), I was called in, and I had no qualms, because I thought that the chances of there being buried explosive there were about a million to one against. There was also a water-diviner, marching about with an impressive assortment of twigs; there were a couple of astrologers, at least one telepath, and various local Saucerers (or Ufologists). The teeming populace was kept away by improvised fences, though I was privileged to go where I liked. The crater was evident enough. It looked as though it had been caused by subsidence, but more than that I could not really say.

We then tried to locate Dr. Randall, but could find only a relative of his who seemed to be a local nurse. Strangely, Woomera disclaimed all knowledge of anyone named Randall; we went so far as to telephone them. So far as I know, nobody has seen him since. (I can add only that he did produce a report on the Charlton affair, saying that on another occasion he had come across a grounded Uranian Saucer and had had a long conversation with the pilot, a gentleman who rejoiced in the name of Ce-fn-x.)

I did track down the rumour that when the space-craft

had landed, it had killed a cow. What had happened was that in a discussion at a local pub, a farmer had said "Ah! and, you know, a cow of mine died last week, too!" with the inevitable result that a journalist overheard, and another sensational headline was dispatched to a newspaper.

I drove past Charlton only a few weeks ago, but all is now quiet there. The craters have presumably been filled in; the crops have grown as of yore; and there have been no further cosmical visitations. But Dorset is not far from Wiltshire, and it is in this admirable county that modern English Saucery has been concentrated of late. The focal point is Cradle Hill, near Warminster, and it was to this eminence that a television film unit and I repaired.

Warminster is a pleasant town, situated in a particularly historic part of England. Like many such places, it has a local paper, and one of the leading journalists on its staff is Arthur Shuttlewood. In addition to reporting mundane matters such as mothers' meetings, football matches and discussions on the Council, he has become extremely interested in flying saucers, and has had some remarkable experiences which he has written down in his several books on the subject. Let me stress that as well as being a good journalist, he is a completely sincere and dedicated person for whom I have the greatest personal respect. I must make this clear, because anyone with my own type of nasty, sceptical mind finds it difficult—nay, impossible—to accept the more extreme parts of modern Saucery. On this point, Arthur and I agree to differ. Meantime, I must become something of a journalist myself, and set out the events which took place on our memorable expedition to Cradle Hill.

We had a long preliminary talk in Warminster itself, and I was able to take in the general background. Arthur confessed that he had at first been sceptical about the whole thing, but after a series of reports from impeccable people such as the head postmaster, the Vicar and a hospital physiotherapist, he decided that Saucers must be taken seriously. It all began, apparently, on Christmas Day, 1964, but then he saw a saucer for himself in September 1965, and had been converted.

My initial question was: "Where do you think they come from?" This was pertinent, even if ungrammatical, and Arthur had the answer to it. Venus was the usual choice, or possibly Mars; but it was significant that most of the Warminster Saucers seemed to come from the constellation of Ursa Major, known officially as the Great Bear and unofficially as the Plough. Now, there are thousands of prehistoric stone circles in Britain, and of these about two hundred are named after King Arthur. All of them are aligned north to south. The Welsh did not have a warrior king comparable with Arthur; but when translated into Welsh the name Arthur means, literally, Great Bear. Therefore it was quite obvious that the Warminster Saucers must come from Ursa Major.

Having cleared this up, I pressed on. "Why Warminster?" This was my next question, and again Arthur (Arthur Shuttlewood, that is to say; not the king) came up with a prompt answer. The root cause is the abnormal magnetic field over this part of Wiltshire. He pointed out that when a military exercise had taken place nearby, in the general direction of the plundered and infamously-seized village of Imber, the signals officers had been completely confused by picking up wireless messages that they certainly did not expect. In fact, they had received signals which had been transmitted during a naval exercise in the English Channel five months earlier, so that because of the low-pressure magnetic field they were, in effect, picking up messages from the past. Flying Saucers have always been regarded as vehicles which can draw their main power from magnetism and surround themselves with force-fields, so that they would logically be drawn to Warminster, where conditions were so exceptional.

As yet, of course, it is impossible to be more specific; but Arthur was confident that the Saucers use a hyper-dimensional link of some sort, which does not fit into our concept of natural science. They may even be able to change their atomic structure, dematerializing and then joining a light-beam of some kind in order to cross tremendous distances in an amazingly brief period. This would make it possible for them to come here from far beyond the Solar System. Those which do not originate in Ursa Major probably come from Venus. True, the surface of Venus is at a temperature of about 900 degrees Fahrenheit, but other life-forms might exist in a higher octave, so to speak, and would also be in a different dimension, totally unaffected by heat which would fry any representative of *Homo sapiens* like an egg.

We then examined various pictures which had been taken by members of the Warminster saucer-spotting group. I commented that several of the objects looked like balls; Arthur agreed that they gave a perfectly spherical appearance when going away at that particular angle. From Cradle Hill they had often been seen coming over, presumably at heights of only a few miles. Attempts to signal them had been made, and on one occasion a space-ship had been seen to drop down no more than a mile away. Most of the visiting craft were friendly, if not all; but efforts to contact them directly were usually unsuccessful. "The faster one runs toward them, the faster they move away," Arthur commented, "I think because of the power-shields, which is still switched on." The record number of sightings in one night was thirteen. This was at Whitsun, 1966. The grand total did not include objects such as high-flying jet aircraft and artificial satellites.

The most interesting part of the whole discussion was Arthur's description of what happened when he had made personal contact with a Saucer—with somewhat alarming results. It had come over the top of the hill, seven or eight hundred yards up; it was of tubular shape—a "flying cigar", at least two hundred feet in length. As soon as Arthur saw it, he dashed into his study, picked up a camera, and prepared to take a photograph. Immediately he felt a tingling sensation along his fingers, upper arm, left eye and jaw; and for some time afterwards his eyes watered and his jaw was partially paralysed. Nonetheless, he took his picture. The space-ship darted off into a cloud, and when next seen was a good three miles up, making its departure in leisurely fashion. When the photograph was developed and printed, nothing came out.

Evidently the power-force in the Saucer was directed toward Arthur, either to stop him firing a gun at it (which, needless to say, would have been the very last thing he would have wanted to do!) or else to stop him from taking a picture. On another occasion, one of the small spheroids literally chased a party of watchers through the copse on Cradle Hill. It was no bigger than a soup-plate, and must have been an automatic beacon sent down from one of the craft to make observations. Then, too, there was the time when a Saucer, coming into the copse from the south-west, produced a perfect arch of brilliant, silvery light, in the midst of which appeared two giant forms: silhouetted figures, long hair waving as though in the wind, with no visible features, but with fingers and robes well defined. For some unexplained reason, the military party down the hill saw absolutely nothing unusual, and it may well have been that the display was put on for one member of the saucer-spotting group, Lady Marjorie Stewart, who had come all the way from Australia to Wiltshire so that she could make an on-the-spot investigation of the Warminster mystery.

By this time I was all agog to go to Cradle Hill myself, and after darkness we duly repaired there, together with the film unit. It proved to be a fascinating evening. The summit of the Hill is a distinctly bleak place, well out of

Warminster, and just the sort of spot where one would expect to meet assorted bogles. Quite frankly, the television lights were a darned nuisance. One cannot gaze upward, looking for a Saucer, when one is being dazzled by a lamp which seems to be stronger than anything perched on top of the Eddystone Rock. After some time, however, I persuaded the team to dim the illuminations, and we began to see luminous specks crossing to and fro among the stars. I thought that they were artificial satellites, but Arthur assured me that they were Saucers, and I concentrated upon trying to train my binoculars upon them. We also saw a glow in the sky, and again there was a difference of opinion; Arthur maintained that it was connected with the interplanetary craft, and rejected my suggestion that it was a low-lying cloud illuminated by the Moon. We stayed for some time, but shortly after twelve o'clock it became clear that nothing more was going to happen, and we decided to call it a day—or, rather, a night.

If Arthur Shuttlewood and his colleagues are right, then what is the purpose of these visitations? I cannot do better than quote Arthur's own words: "I think it will lead eventually to man on Earth becoming a truly civilized creature." With that wish nobody would quarrel—least of all myself.

In a way, I believe that Warminster represented the high water-mark of British Saucery. Nothing of note seems to have happened since, though reports have continued to come in, and the various societies and clubs devoted to the subject meet regularly and issue their pronouncements. Come what may, historians of the future will find the whole episode vastly intriguing. As I have said, no other Independent Thinkers of modern times have made comparable impact. There have been official Government inquiries, innumerable books, countless news items and myriads of published letters. What would really settle the matter, of course, would be a visit from an interplanetary traveller

107

who would be prepared to make himself known to the world at large rather than do no more than flit around, pausing now and then to chat to people such as George Adamski and Cedric Allingham.

I must be honest, and say that I have no real hope that this will happen. But if it did, I would welcome it, because even a casual glance at the newspaper headlines on any typical day in A.D. 1972 will show how much we Earthmen need guidance from some exalted being who knows more than we do! Moreover, I admit to having an ulterior motive. There is nothing I would enjoy more than introducing a Martian, a Venusian or even an Alpha Centaurian to viewers of my own television programme, *The Sky at Night*. Should any Saucer pilot happen to be reading this book, I hereby extend him a cordial invitation to drop me a postcard as soon as he has a spare moment in his busy programme.

11

REPORT FROM MARS, SECTOR 6

Walk down the Fulham Road, threading your way through London's traffic, and you will eventually come to No. 757. It is a remarkable place. Here is the headquarters of some of the world's most surprising Independent Thinkers—the members of the Aetherius Society. It is part-scientific, part-religious; according to its publications, it has been charged with a mission of vital importance to mankind. Its founder, George King, has been selected to become Earth representative of the Interplanetary Parliament, which, of course, meets on the planet Saturn.

To show the fundamental nature of its mission, I may cite two examples, both of which have been described more fully in the Society's publications (of which the most notable is the periodical *Cosmic Voice*). First, there was the crisis of the 1950s. Unfriendly fish-men from the far side of the Galaxy launched a missile at us, with the set intention of removing us from the face of the universe. This was because the fish-men lived on a planet which was drying

109

up, and they considered it wholly justifiable to land on Earth and take possession of our oceans. Fortunately, the news filtered through to the Interplanetary Parliament, and it was decided—we hope unanimously—to take strong action. The Master Aetherius, who lives on Venus and has special responsibility for terrestrial affairs, was away at the time, but the Martians were on the alert. With commendable efficiency they launched a thunderbolt at the missile, and destroyed it. It is not now thought that there will be another attack in the foreseeable future; but without the Aetherius Society, we would never have known that we had been under so dire a threat.

Next, between January 20 and February 9 in 1962, came another time of trouble. At that period several of the planets were in the same part of the sky, so that they were pulling in roughly the same direction. The astrologers were not the only people to be worried by this. The days of greatest concentration (February 4 and 5) were the subject of special transmissions from Aetherius, sent, as usual, by way of Mr. King. These transmissions, beamed from the giant planet Jupiter, were sufficient to counteract the modifications which Man had imposed upon the radiations coming from afar. However, everything depended on the members of the Aetherius Society, who, "while millions slept in blissful but dangerous ignorance", tuned in to Jupiter to be used as channels for the great Transmitting Energies. Eventually the danger subsided; the power could be switched off, and all was well once more. As was stated in *Cosmic Voice*, Jupiter had saved Earth, and had used the Aetherius Society personnel to do so.

When I heard about these and other events, I was naturally very grateful, and it was clear to me that no book about Independent Thinkers would be even remotely complete without introducing the Society. By this time Mr. King had emigrated to America, to found a new branch of the Society in Hollywood; he had also become first Dr. George King, then the Rev. George King, and then the Rev. Dr. George King. (He still is.) However, the London branch of the Society was most courteous and co-operative, and Mr. Robertson, one of its senior members, kindly acted as spokesman. The rest of the information in this chapter is drawn partly from what Mr. Robertson told me and partly from the many publications which have been issued since May 1954, when Mr. (I am sorry: the Rev. Dr.) King received the first intimation that he was to become the Voice of Interplanetary Parliament.

He was not entirely prepared for this sudden responsibility, but he was well equipped to meet it. For the previous ten years he had been a keen student of Yoga and allied subjects, and was well versed in Metaphysics. Every moment of his spare time was devoted to this research. (I believe that he was then officially a London taxi-driver; this may not be correct, but at any rate he was doing a thoroughly useful and honourable, albeit mundane, job.) However, the message which came to him on that fateful morning in May was quite straightforward, especially as it was delivered in perfect English. Mr. King was told, boldly: "Prepare yourself. You are to become the Voice of Interplanetary Parliament." A mere eight days later he was visited by a Master of Yoga, who walked in to see him, taking no notice whatsoever of the locked door which happened to be in the way. After the conversation, Mr. King made ready for direct contact with intelligence from space. And a few months later, the Rev. Dr. King had his first personal communication from a Cosmic Master who lives on Venus and who uses the pseudonym of Aetherius.

The die was cast. George King accepted the challenge. In January 1955 he gave his first public address, at Caxton Hall in London, after which he threw himself into a trance. This allowed Aetherius to come through in person, all the way from Venus. He has, of course, the advantage of being able to speak all terrestrial languages—apart, so

111

far as I can gather, from French and Norwegian.

Meantime there had been great activity out in space. It is not generally known that there are usually at least two artificial satellites orbiting the Earth; not our own, but dispatched from afar with the full authority of the Interplanetary Parliament. Of these, the most important is the Third Satellite, manned by experts able to manipulate huge potential energies for the benefit of mankind. While the satellite is operating, we pass through what are called Magnetization Periods or Spiritual Pushes. The whole sequence of events is supervised by an advanced Martian scientist who is in regular touch with Dr. King, and is referred to as "Mars, Sector 6". Much of the information which follows is drawn from this source.

We have detailed information about the Third Satellite, because Dr. King has been there. In the sixth issue of *Cosmic Voice*, the periodical which he founded at an early stage after his election as Earth Representative, he gave a long account of his trip, which happened on March 23 1956. The Satellite is a remarkable place. It is oval, and bathed in its own coloured lights. The main room is scrupulously clean, and filled with an alluring perfume; in other words, it smells. There is a large glass window in the ceiling, allowing only selected radiations from space to enter; these radiations affect the giant prismatic crystals inside the craft, with spectacularly beautiful results. Dr. King learned that the scientists have been able to isolate the "universal life forces" by the simple process of switching coloured lights on and off. The resulting energy can be directed to any spot on Earth.

Dr. King did not meet Mars, Sector 6 in person; but he did meet one of his colleagues, who looked about thirty, but was in fact three hundred years old. Like all good Cosmic Adepts, he had long hair reaching down to his shoulders (let me repeat, this was well before the era of student yobs), and he wore the usual kind of natty one-piece space-suit.

Subsequently there were other contacts, mainly by Dr. King himself but also by some of his followers, so that by August 1956 it had become possible to found the Aetherius Society. Under its inspired Chairman-Founder it thrived, and today it has branches in many countries, including the United States. Indeed, for some years now Dr. King has found that his great duties can best be performed by operating from Hollywood, California, rather than the Fulham Road—an inconvenience which he has borne with his customary fortitude.

The original Articles of the Society were laid down by an important personage named Saint Goo-Ling, about whom I have no reliable information. It was stressed that no other Society had ever had to deal with tasks of such importance, and for this reason alone I was impelled to go to one of the early meetings at Caxton Hall, London. It must have been about 1957; I still have my notes of the meeting, though they are unfortunately undated. (I do remember that we were still in the pre-Space Age; that is to say, before the launching of Russia's first artificial satellite, Sputnik I, on October 4 1957.) Following the address, given personally by Aetherius by way of Dr. King's vocal chords, questions were called for. I was ill-natured enough to ask one in Norwegian, which caused bafflement. I then repeated it in French, but this was no good either, and finally it had to be put in English. What I wanted to know, quite simply, was: Where *was* Aetherius at that moment? On Venus, in space, or smiling benignly down on us from the rafters, invisible to our mortal eyes? Dr. King was unfortunately unable to answer. I was merely told, rather brusquely, that Aetherius' present whereabouts was a secret.

Incidentally, Aetherius is by no means the only Cosmic Master in touch with the Society. Another is no less a person than Jesus Christ, who is Venusian by birth and who originally visited our world in the flying saucer which is referred to in the Bible as the Star of Bethlehem. At present he lives on Mars, and he was personally involved during the little contretemps with the fish-men referred to above (it was lucky for all of us that he had kept his eyes wide open). The fact that both Mars and Venus are unpleasant worlds judged by our own standards is quite irrelevant, because the Adepts function according to entirely different vibrations, and do not need our type of environment. Neither do they need to eat. Their intake of energy is merely by breath, so that, for instance, anyone on Saturn can presumably gulp down a few lungfuls of ammonia or methane and then feel as refreshed as Popeye after a complete tin of spinach. Aetherius, we are told in the Society's official booklet about flying saucers, has not eaten for over three thousand years, and as yet there is no indication that he feels even slightly peckish.

I mention Jesus Christ at this juncture because it was he who initiated another of the Society's activities. It was on July 23 1958, at Holdstone Down, a hill in North Devonshire, England. Dr. King had been summoned there (direct orders from Aetherius; a sort of cosmical three-line whip) and was told that it was necessary to charge various mountains with cosmic energy, to act as relays for Spiritual Pushes. Of course this command could not be ignored, and the Society went into action. Before the end of February 1959 they had laboured to the tops of nine British mountains, and charged them with sizzling energy. Dr. King then undertook a world tour to carry on the good work, charging mountains as far afield as Australia and New Zealand before coming back home. The last mountain to be treated was Mont Blanc. Aetherius was well pleased, and even said that the operation was "the most important single metaphysical task ever undertaken upon *Terra* in her present life". (*Terra* is Venusian for "Earth". The Romans also used the term.)

Before going any further, I should perhaps give a brief

account of a curious episode which took place in 1957. It is not really part of the story, but I can hardly omit it, because it involved a meeting between Dr. King and myself. By this time *Cosmic Voice* had graduated from a duplicated sheet to a properly printed magazine, and it has, of course, many contributors. Among them was Dr. Dominic Fidler, M.A., D.Sc., who, in No. 10 of the periodical, wrote about "Mescalin and Flying Saucers". This contained a quote from the Swedish scientist, Professor Huttle-Glank, that volcanoes in Mexico could act as temporary power-zones, and could therefore expect to be visited by space-craft. Dr. Fidler's article also contained the suggestion that an observer's perception of flying saucers could be sharpened by the use of drugs such as oxfordine. (Again at the risk of repetition, let me point out that at this time the drug menace in society had not assumed the proportions that it has now; it was still officially condemned, except of course by some brands of psychiatrists.) In the next issue Dr. A. Wells took up this problem, and described some practical experiments carried out by Dr. Pullar and Professor N. Ormuss. And in No. 12 of *Cosmic Voice*, one of the most fascinating ever produced, Professor Huttle-Glank pointed out a mistake in Dr. Wells' article. Oxfordine, he said, was a derivative of gauxine, with percentages of boltzine and haledin; a previous formula, involving habenzine, had been rejected as unsafe. This had been made clear by two Dutch experimenters, Houla and Huizenaas.

In the same issue there appeared an article by Dr. Walter Wümpe, an eminent Austrian scientist, about the Congress on Vibrations in Vienna. Apparently one of the leading papers was by a Leipzig scientist, Egon Spünraas, who had made a vibratory recorder to prove that some of the messages from space emanated from the Star Beta Leonis, 43 light-years away. (One light-year is equal to almost 6 million million miles.) Others who took part in the discussion reported by Dr. Wümpe were Drs. E. Ratic,

Kreme, and Hotère. The Italian physicist Lupi missed the meeting, as he boarded the wrong aircraft at Rome and went to Oslo instead of Vienna. Dr. Wümpe's paper was ably translated by his secretary, R. T. Fischall.

This was all quite in order—but in the meantime the Society had been under fire from *Psychic News*, a paper devoted to Spiritualism. Dr. King challenged *Psychic News* to reprint any article from *Cosmic Voice*. They did. They chose Dr. Wümpe's, suggesting that some of the names and ideas were a little unusual, and ending with the pertinent question: "Huizenaas?"

Dr. King came to the conclusion that some of his contributors were not quite so serious or so scientific as he had been led to expect, and, breathing fire and slaughter, he set out to trace Dr. Wümpe, Egon Spünraas, R. T. Fischall and the rest, not forgetting N. Ormuss. For some reason that I cannot explain, he suspected that I might be implicated, and he came to see me at my home at East Grinstead. I was unfortunately of little help to him, and the whole episode is now so far in the past that we had better skip a decade or so and take up the story again in 1969, when I went to the Society's headquarters in the Fulham Road and had a long, very pleasant and cordial interview with the official spokesman, Mr. Robertson.

After some general introductary chit-chat, I asked whether there was any information about the status of *Homo sapiens*; had we ever lived anywhere else? Mr. Robertson confirmed that many thousands of years ago, the human race was domiciled on a planet which orbited the Sun between the paths of Mars and Jupiter. Alas, our remote ancestors did more than dabble in science. They allowed it to get out of control, and the result was a most unedifying explosion which blew up not only themselves but also their planet. The shattered remains are, of course, the asteroids. Survivors of the blast managed to reach Earth, but we remain in a very primitive state compared with other races in the Solar System.

At this point it may be as well to give a brief résumé of conditions elsewhere in the Sun's family—beginning with Mars, because this planet is the base for most of the flying saucers (the rest come from Venus and beyond). Again we have to depend mainly upon information received from Mars, Sector 6, which normally only Dr. King can receive. (There is also Mars, Sector 8, which is situated on Phobos, the inner of Mars' two tiny moons. Phobos is artificial, and acts as a centre for weather control and communications, though judging from the photographs of it obtained in December 1971 by the American Mariner probe, the installations have been well camouflaged. Sector 8 seems to be much less powerful than Sector 6; possibly its batteries need re-charging.)

It is fascinating to find that the Martians, most of whom live underground, are of terrestrial origin! Just before the destruction of Atlantis, vast space-ships landed to take the Atlanteans to the safety of Mars, where they remain to this day. This is comforting, since when we make general contacts we know that we are dealing with people of the same basic type as ourselves.

Actually, Mars is not too infertile. The canals are strips of vegetation, planted in a way so as to be in complete harmony with the prevailing magnetic fields. Quite apart from its plants, Mars is the industrial planet of the Solar System—a sort of cosmical Sheffield. There are immense Saucer factories, and demands from the Interplanetary Parliament ensure that work is going on all the time. All manufacture is controlled by thought-directed robots. Aetherius has said nothing about restrictive practices, victimization and strikes, so we may safely assume that Trade Unions have not yet been established there.

Teleportation is the order of the day. This, as is well known, involves the transfer of solid bodies through other solid bodies, so that locked doors are futile. Also, a Martian engineer can sit inside his subterranean laboratory and, by sheer thought, direct a vehicle to the far side of the Milky Way, landing it on a desolate world and bringing it home after the exploration has been completed. Eating is superfluous, and there is a highly efficient educational system. Schools are built of crystal, so that the young people may be impregnated with magnetic vibrations coming from other parts of the Solar System. I have not been able to ascertain whether there is a Martian equivalent of the British G.C.E. examination; but I am sure that if their methods were introducd here, they would work much better than those now being practised in Britain's Secondary Modern and Comprehensive schools.

Moving outward we come to Jupiter, which is the "reception planet" of the Solar System. Space-travellers coming from afar always book in there first. Again most people live underground, but there is no objection to residing on the surface, or indeed upon one of the natural satellites, of which there are twelve. Music is very important to the Jovians. Indeed, Music Energies are regularly dispatched to Mars in exchange for flying saucers, which seems to be an arrangement suitable to all concerned.

Deferring consideration of Saturn for the moment, we must pause to mention Venus, about which we have particularly complete information inasmuch as Aetherius himself lives there. (Among other prominent Venusians are Buddha and, as we have seen, Jesus Christ; Confucius, however, was a Saturnian—at least, so he say.) Venus is an extremely advanced planet, with a magnificent Crystal Temple which radiates immense power during a Spiritual Push. According to the descriptions given, a typical Venusian is in the region of seven to nine feet tall, with long white or fair hair, a cinnamon-coloured skin, and eyes without pupils. Their feet seem abnormally small to carry so tall a body. It is possible that Aetherius looks like this, though no specific description of him has been sent; we must simply wait until he decides to show himself to us, as will no doubt happen in the foreseeable future. There is certainly no transport problem. A Venusian Saucer can cover the distance between its home base and Earth in a mere two and a half seconds. (British Rail, please copy.)

Some of the Saucers are very large, and a "mother-ship", capable of carrying up to 7,000 Scout ships, may be as much as 5,000 miles in length. Each is propelled by a magnetic device which exerts an equal thrust on every atom of substance, thereby cancelling out the effects of gravitation. Apparently all this is perfectly simple when one really puts one's mind to it. The orbiting satellites, of which Mars, Sector 6 is the most potent, are decidedly smaller; they are absolutely invisible, and cannot be detected even by radar.

What, next, of our own puny efforts in space?

In August 1964, Aetherius announced that in future terrestrial space-ships would be allowed to land upon the Moon; to make this easier, the Venusian, Martian and Saturnian bases which had been set up there were dismantled. Things had not always been so easy. For instance, in 1960 a Russian lunar probe was actually destroyed by magnetic beam Heron Six Five; this was categorically stated in a message from Mars, Sector 8. However, since then the Apollo astronauts have made their explorations unhindered, and at the moment the Moon seems to be uninhabited. Travel to Mars by manned craft is still not allowed, but probably the restrictions will be lifted as soon as we have learned enough to send expeditions there. At last, we can justifiably hope so.

Next, and perhaps most important of all, what about the Interplanetary Parliament? Since this is located on Saturn, I asked Mr. Robertson what the Saturnians looked like. He replied that by this stage in its evolution, a Being has gained such a status that it can be seen in its entirety

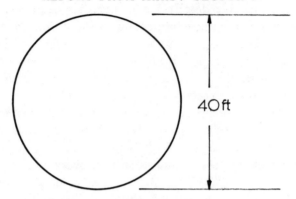

Fig 11   An inhabitant of Saturn, according to Aetherius

—that is to say, an ovoid shape: a great incandescent egg perhaps forty feet tall. I commented that the Saturnians seemed to be extremely large balls, and Mr. Robertson agreed that this was a perfectly sound description.

Saturn itself is the Home of Wisdom. On its surface there stands the Temple of the Ancient Ones, made of crystal and shining as brillantly as the Sun itself. Inside it are the Twelve Perfect Beings. There they stay, immobile, and wrapped for millions of years in their wonderful Silence. No rowdy parties, pop groups or protest marches belong here. In fact, so far as I could tell, nothing happens at all. It may be rather dull; but only a Saturnian knows.

By this time I was almost speechless myself, so I asked what would be the future of the Aetherius Society. Mr. Robertson replied that it would continue as a centre for Space Intelligence; and after the end of Dr. King's own time on Earth (which, we all hope, will be delayed for many years), methods will be found for continuing the transmissions from Mars, Sector 6 and elsewhere.

I do not really think that there is much to be added. In my view, at least, the members of the Aetherius Society are the most astounding of all the Independent Thinkers. They are absolutely sincere; they have a large following;

and as the years go by they will continue charging their mountains, reporting Spiritual Pushes, and relaying the cosmic news via Dr. King and any other chosen Voice of the Interplanetary Parliament. They are quite convinced that the world's problems are much less likely to be solved in Downing Street, the White House or the Kremlin than in the Fulham Road. I am only sorry that I myself can have no real hope that I will be able to travel upward to the Temple of Silence and gaze, awestruck, at the forty-foot incandescent eggs which rule the destinies of us puny mortals.

12

PHASE THREE OF SPACE RACE

This will be an extremely short chapter. The material contained in it certainly merits inclusion in the present book, but I did not know quite where to put it (please do not misunderstand me).

I have tried my best to get in touch with Mr. Theodore B. Dufur of the United States, who in 1955 had proposed an edible space-ship to be made chiefly of substances such as frozen oleo margarine. His plan was simple enough, and was widely publicized. All you do is to fly to the Moon and then nibble away at your space-craft pending the time when the next expedition arrives. Things are even pleasanter if you land inside a lunar crater in which volcanic activity is going on; you can drop a lid over the crater and keep yourself warm.

Unfortunately I was unable to track down Mr. Dufur, and neither did I have any success with the Orion Observatory in Santa Monica, California, which had published a paper showing that some of the lunar rocks shown on

Fig 12 Diagram of a lunar Kritosaurus, traced from a photograph
of the Moon's surface

the photographs taken on the Moon by the automatic
Surveyor probes were nothing more nor less than the
remains of dinosaurs. My third failure was with Mr.
Edward Mukaka Nkoloso, to whom I now turn.

It would be quite wrong to suppose, as many people do,
that America and Russia are the only nations involved in
sending probes to the Moon and Mars. Zambia, too, is in
this field of research, and the leading Zambian astronauti-
cal planner, Mr. Nkoloso, undoubtedly qualifies as an
Independent Thinker inasmuch as his ideas about space-
travel differ so widely from those held in Houston or
Moscow.

I first heard of him from an official news report on
November 3, 1964, which I quote as follows:

"America and Russia may lose the race to the Moon,
according to Edward Mukaka Nkoloso, Director-General
of the Zambia National Academy of Space Research. His
ten Zambian astronauts and a seventeen-year-old African
girl are poised for the countdown. He said: 'I'll have my
first Zambian astronaut on the Moon by 1965. My space-
men are ready, but we're having a few difficulties...we
are using my own firing system, derived from the catapult."

"Mr. Nkoloso continued: 'To really get going we need
about seven hundred million pounds. It sounds a lot of
money, but imagine the prestige value it would earn for

Zambia! But I've had trouble with my space-men and space-women. They won't concentrate on space-flight; there's too much love-making when they should be studying the Moon. Matha Mwamba, the seventeen-year-old girl who has been chosen to be the first coloured woman on Mars, has also to feed her ten cats, who will be her companions on the long space-flight.... I'm getting them acclimatised to space-travel by placing them in my space-capsule every day. It's a 40-gallon oil drum in which they sit, and I then roll them down a hill. This gives them the feeling of rushing through space. I also make them swing from the end of a long rope. When they reach the highest point, I cut the rope—this produces the feeling of free fall.'"

For some reason that I cannot explain, the United Nations authorities turned down the Z.N.A.S.R. request for £700 million, and it was not until December 29 1968 that Mr. Nkoloso again hit the headlines. In another official dispatch, he congratulated the successful Apollo 8 astronauts who had just completed the first manned flight round the Moon. He continued with a clarion call: "Let us make a Zambian rocket today. We shall never be content to remain behind other races. This is our heavenly destiny, our natural ambition and our scientific and cultural hegemony." He added that the Americans had parsimoniously refused him the paltry twenty-one million dollars he had wanted for the purchase of liquid oxygen, and he maintained that it was only the lack of finance which was keeping his space programme on the ground. Meanwhile, his astronauts were keeping on with their training—cats and all.

There followed another lull, but on August 20 1971 Mr. Nkoloso announced that the Zambia Space Research Academy was constructing a new telescope "to see the planets clearly". Mr. Nkoloso said that the telescope would be on display at Chunga Valley, ten miles west of Lusaka, and that two technicians and three engineers had been working on it for three months. It was to be officially opened in September with a display by twelve Zambian astronauts.

I was intrigued (well, wouldn't you have been?) and wrote to see whether I could obtain an invitation to the grand opening. However, I had a letter from the Ministry of Technology in Lusaka saying that the opening had been postponed; and up to the present time I have heard no more.

Life is full of these disappointments. But I continue to hope.

13

FROM CHARIOTS TO PYRAMIDS

On June 30, 1908, a funny thing happened on the way to Tunguska, in Siberia. Without the slightest warning a brilliant light appeared in the sky; an object of some kind shot gaily across the heavens, and landed with a thud which was recorded by seismometers many hundreds of miles away. A column of fire and smoke shot upward, and pine-trees were blown down like so many matchsticks. For fifteen miles around the point of impact the devastation was more or less complete, but fortunately nobody was killed, for the excellent reason that nobody lived there.

Scientists came to the very reasonable conclusion that the Siberian tundra had been struck by a meteorite. There seemed nothing improbable about this. Meteorites are not uncommon; most museums have collections of them, and sometimes we come across objects of large size. The holder of the heavyweight record is the Hoba West Meteorite, in Africa, which exceeds sixty tons, and is still lying where it fell in prehistoric times. (I doubt whether anyone will try to run away with it.) In Greenland, the explorer Peary found a 36-ton meteorite which he duly scooped up and brought home; go to the Hayden Planetarium in New York, and you can see it for yourself.

This being so, the Siberian object was dismissed as being nothing more than an exceptionally large meteorite. No expedition to the site was dispatched for some time, because things in Russia were in a somewhat unsettled state; and after the Red Revolution, some years later, the local Bolsheviks were too intent upon blowing each other up to worry much about meteorites. When a party finally did reach the area, in 1927, traces of the explosion were still very obvious. A Russian astronomer, K. Florensky, later suggested that instead of being a normal meteorite, the object might have been the nucleus of a small comet. This again is quite rational, but other theories too have been proposed, and this brings us on to Comrades Kazantsev, Altov and Shuraleva.

Alexander Kazantsev, of Moscow, is a science-fiction writer who occasionally puts forward theories of his own. I have never met him; I first heard of him when I read his paper about the Siberian Meteorite. In brief, he suggested that it was nothing more or less than a space-ship which had come to visit Earth from some remote planet. It made a faulty approach; the atomic motors exploded, and instead of making a gentle landing the space-ship blew itself and its crew into a great many pieces.

This idea is admittedly hard to check. Contrary to some earlier reports, there are no signs of latent radioactivity in the area, and neither is there any evidence that the falling object may have been composed of "anti-matter", which would go up with a loud bang on contact with ordinary matter of the kind which makes up ourselves. All one can really say it that if Kazantsev were right, the original space-ship designers would have to go back to the drawing-board.

Even more unusual was the idea proposed in a Leningrad paper, *Svesda*, in March 1964. The authors were G. Altov and V. Shuraleva, neither of whom had achieved world-wide renown, but both of whom were clearly prepared to be controversial.

When discussing Atlantis, I had occasion to mention the Krakatoa eruption of 1883, the greatest natural outburst of modern times. To refresh your memories: a fair-sized island was blown up, huge tidal waves devastated the coasts of Java and Sumatra, and the dust stayed in the upper atmosphere for three years. Anyone watching the sequence of events from a vantage point out in space would have seen a very considerable disturbance. Even the most myopic Saucer pilot would have realized that something unusual was going on.

You may well be wondering what Krakatoa has to do with the Siberian explosion. According to Altov and Shuraleva, the connection is very close. They believe that a long way away from us, circling a star in the constellation of Cygnus, there is an inhabited planet whose astronomers take a lively interest in Earth. On seeing the Krakatoa outbreak, the Cygnians (if we may so call them) jumped to the conclusion that we were trying to send a signal. This, of course, could not possibly go unanswered, and so the ingenious Cygnians replied by turning an energy-beam toward us. Unfortunately they miscalculated. The beam was much too powerful, and when it hit the top of the Earth's atmosphere it fused. The energy was changed into a chunk of matter—and it was this piece of material which shot downward, finally coming to an unceremonious halt in the Siberian pine-forest.

If Krakatoa blew its top in 1883, and the Cygnians did not see it until 25 years later, then presumably they are 25 light-years away from us. We can envisage setting up an interesting series of experimental messages. According to our Russian colleagues, a volcanic eruption is answered by a death-ray. We might possibly reply by triggering off an earthquake, which would elicit a response in the form of a well-aimed comet; and so on. However, we must be cautious. After 1908, successive 25-year intervals would bring us to 1933, 1958 and 1983. So far as I can remember, nothing untoward happened in the first two of these critical years; but if we have another major "meteorite" in 1983, we may have to do some hard thinking.

The idea of sending deliberate signals to other worlds is far from new. Way back in the first years of the nineteenth century, plans were put forward for digging geometrical patterns in the Sahara Desert, so that astronomers on Mars would see them and reply suitably. There was also a remarkable Frenchman, Charles Cros, who was active in the 1870s. He wanted to build an immense burning-glass, capable of focusing the Sun's rays on to Mars and making scorch-marks on the surface of the planet; once this had been achieved, it would be possible to swing the burning-glass round and so write words in the Martian deserts. For some reason or other the French Government was not enthusiastic; and though Cros continued his propaganda for some time, the scheme was never actually tried.

(*En passant*, I wonder what words Cros proposed to write? And more pertinently, what would our own reactions be if large letters started appearing in the Sahara? Quick-fire repartee would be out of the question, even if the message-senders were considerate enough to transmit in the usual excellent broken English.)

Before leaving the Soviet Union, I must say a brief word about Dr. Iosif Shklovskii, who is undoubtedly one of the world's leading astrophysicists. As we have noted, Shklovskii believed Phobos and Deimos, the two miniature satellites of Mars, to be artificial space-stations. He came to this conclusion mainly as a result of certain alleged irregularities in the movements of Phobos. The calculations have since been found to be wrong, and in any case the Mariner 9 pictures of Phobos, taken in December 1971, show a body which looks decidedly natural.

It is quite true that in 1726 Jonathan Swift, author of *Gulliver's Travels*, described how the astronomers of his curious flying island, Laputa, had discovered two dwarf moons of Mars, both very close to the planet. Flying Saucerers and Atlanteans have been quick to claim that Swift knew all about the satellites, even though during his lifetime there was no telescope powerful enough to show them. Had he access to knowledge locked up in the Ancient Teachings? Perhaps; but another explanation occurs to me. In Swift's time the planet Venus was thought to have no satellites; the Earth had one, and Jupiter four. Therefore Mars, coming in between Earth and Jupiter in order of distance from the Sun, might reasonably be expected to have two moons. As was said at the time, how could Mars possibly manage with less?

The Ancient Teachings lead us on to another branch of Independent Thought: the theory that the Earth was visited by spacemen in the remote past, and that even we ourselves may be descended from them. One complication is that the various suggestions are tangled up with Atlantis, Lemuria, Mu and even the Bible. The Book of Ezekiel begins with a remarkable account of what has been regarded as a space-ship carrying four astronauts, each of which had four faces and four wings. Fiery wheels were also very much in evidence. It must have been a most upsetting spectacle, and one can hardly be surprised that Ezekiel fell upon his face. Frankly, I think I would have done the same. He was not entirely reassured when he found himself being spoken to, presumably by one or all of the astronauts.

I mention Ezekiel here not only because he figures in the Bible (and remember that according to Mars, Sector 6, Jesus Christ himself was a Venusian) but because he leads us on to Mr. Erich von Daniken, whose book, *Chariots of the Gods*, has had a vogue rivalling those of Adamski and Velikovsky. I have yet to meet him, as at the time of my television programme on the subject he was unable to join me, as he was detained in Switzerland. He is quite sure that the fiery wheel of Ezekiel was a space-ship, and that God also was an astronaut. Indeed, it is likely that in Biblical times—that is to say, around 2,000 years ago—there were quite a number of visitations from space. When the accounts came to be written down, centuries later, the various astronauts were unceremoniously lumped together and called God.

Let us admit, without reservation, that so far as chronology is concerned there are many gaps both in Biblical and in pre-Biblical records. If space-visitors had once made a habit of dropping in on us, later writers would have found it impossible to pinpoint their comings and goings at all accurately. In support of his hypothesis, Mr. von Daniken goes back a long way into the past—in fact, to the period when men lived in caves. He believes that some

Fig 13 A cave-sketch of an astronaut

examples of cave art show what appear to be helmeted astronauts. This may, I suppose, be the case, but one is bound to be rather dubious. There are many people (including myself) who would find it hard to draw a human figure recognizably even with the aid of modern pencil and paper. Our cave-man artist, chipping out his work on stone with the prehistoric equivalent of a chisel, would be at an initial disadvantage. Many years ago, I remember being shown some grotesque seventeenth-century carved heads on display at an art exhibition. One of them was remarkably like an aunt of mine; but I did not seriously think that my aunt was on this earth three hundred years ago, even in a former incarnation.

Quite apart from the cave-men, Mr. von Daniken cites cases of knowledge which was current in ancient times, and yet which could hardly have been gained without strictly modern equipment. In particular, there are the maps in an atlas compiled by a sixteenth-century Turkish cartographer, Admiral Piri Reis. To me they do not seem very clear, but according to Mr. von Daniken they show not only the Americas, but also Antarctica, with amazing precision. Transferring them to a globe, it is found that they can only have been compiled from photographs taken from a space-ship which was hovering at a great height dead above Cairo. And in Peru, near the ancient city of Nazca, there are gigantic linear patterns which, it is said, mark the remains of an ancient airfield from which Atlantean-type vimanas used to take off and land. Mr. von Daniken adds that the time-scale agrees excellently with that required by Herr Hörbiger's WEL.

Following up these ideas, we come on to the notion that mankind may be the result of a genetic experiment carried out by visiting astronauts. According to this theory, the story began in the very remote past, when an unknown space-ship arrived on Earth (possibly to refuel) and found primitive life-forms here. No doubt from a spirit of pure scientific inquiry, the space-men fertilized some female members of the primitive Earth species, and went away well satisfied. Much later they came back, and found that their efforts had resulted in a general increase in local intelligence. The fertilization process was repeated—perhaps several times, at suitable intervals—until, at last, the Earth creatures became brainy enough to start drawing on cave-walls. Rather obviously, they were full of awe and respect for their visitors. Hence the graphic sketches of helmeted astronauts....

Mr. von Daniken does, however, concede that the fertilization theory is "full of holes", and adds that the future will show how many of these holes can be filled in. I do not feel that I can add anything useful to this pungent comment.

Whether or not you and I are descended from Martians or Venusians (or Cygnians, or even Alpha Centaurians), we must agree that there are some features of the ancient world which need a good deal of explaining away. There are the strange statues in Easter Island, which, it has been suggested, were fashioned by visiting astronauts who were in a mood of cheerful relaxation after their Herculean efforts to improve the quality of the human species. But of paramount importance are the Egyptian Pyramids, which, according to Mr. von Daniken and others, could hardly have been built by conventional methods known at the time. This provides us with a convenient link between space-visitations and Pyramidology—which still has its followers, even though it is much less popular now than it used to be a century ago.

One of the regrets in my life is that I have never seen the Pyramids. They must be impressive by any standards, and they were undoubtedly of great significance to the Egyptians who built them. The Great Pyramid of Cheops remains unique; it cannot have been planned or erected quickly, and even with the aid of 1972-type machinery it would be a major undertaking. As a friend of mine picturesquely said after having been to it, there is an awful

133

lot of Pyramid. It is almost five hundred feet high, and it is said to weigh rather over thirty million tons. Mr. von Daniken speculates as to whether it and the other Pyramids were built by order of the astronauts, and under their supervision, so that leading personalities could be "deep-frozen" and stored there in between the cosmical visitations.

All high-class space-men know the secret of anti-gravity, and this would make pyramid-building relatively easy. You merely select your block of material, make it weightless by waving a wand over it (or by reciting some incantation, such as "Abracadabra"), and beckon it along to its selected site. You then switch off the fluence, allowing the block to settle cosily down. After that, suitable chipping and fashioning will turn the shapeless block into a proper pyramid.

Even if these speculations are wrong, and the Pyramids were constructed purely by denizens of Earth without help from outside, we still have a great deal to interest us. Modern Pyramidology was born in 1859, when a London publisher, John Taylor, wrote a full account of the Great Pyramid—admittedly without ever having been to it. He was convinced that it had been set up by the Israelites, under the direct guidance of God. He also believed that its dimensions were linked with those of the Earth, the Solar System, and indeed the entire universe.

Taylor was followed by Charles Piazzi Smyth, Astronomer Royal for Scotland, who had quite justifiably earned a great reputation as a skilled and energetic astronomer. Smyth not only accepted all that Taylor had written, but did his best to improve upon it; and unlike Taylor he went to Egypt to make an on-the-spot investigation. Eventually he published a book, *Our Inheritance in the Great Pyramid*, which is a classic of its kind, worthy to rank with Velikovsky and Adamski at their best.

Smyth made vast numbers of measurements of the Pyramid, and drew startling conclusions from them. Let me give a few examples. The base of the Pyramid, divided by the width of a casing stone, equals 365: the number of days in a year. Each casing stone is slightly over 25 inches wide: this is precisely one ten-millionth of the Earth's radius as measured through the poles. Multiply the height of the Pyramid by 10^9, and you obtain a distance which is about equal to that between the Earth and the Sun. Clearly, then, the Pyramid was built by people who knew all about the correct scale of the Solar System—which the conventional Egyptian astronomers certainly did not; they believed the world to be rectangular, with Egypt in the middle.

Of course, things were not *quite* right. The casing stones measured a little over 25 inches each, and to round things off neatly Smyth devised the "pyramid inch", a sacred unit slightly shorter than the normal inch. As soon as this great discovery had been made, all important past and future events could be worked out, merely by measuring details of the Pyramid. For instance, if you assume one pyramid inch to represent one year, the length of one of the main passages—4004 pyramid inches—proves that the world began in B.C. 4004, as postulated by Archbishop Ussher of Armagh. Another measurement indicated that Christ would return to Earth at some date between 1882 and 1911. This seemed reasonable enough in 1865, when Smyth wrote his book, but, so far as I know, the prediction was not fulfilled.

Unkind people claim that by suitable tinkering with figures it is possible to prove almost anything, and there is a well-authenticated story of an earnest Pyramidologist who spent some time in filing down a casing stone because it was slightly too long to fit into the general plan. To test the mathematical theory, I went to see an old friend of mine, Henry Brinton, who lives a few hundred yards away from me in Selsey. We made some measurements of his home, Old Mill House, and proved from them that men would certainly reach the Moon in 1969. You will have to accept my word that this experiment was carried out in 1964, when the first lunar landing was still some time away!

During the present century, Pyramidology seems to have declined, and when I cast around for an enthusiast to discuss the matter with me, I met with utter failure. Perhaps the truth is that nobody could possibly hope to out-Smyth Smyth.

I am afraid this has been a somewhat rambling chapter, but this is inevitable when one is delving so far into the past. Whether any new evidence will turn up remains to be seen. I suppose it is not out of the question that if Mr. von Daniken is right, the astronauts who were originally responsible for creating mankind—and the Pyramids—will come back one day to see how we are getting on; but if so, I hardly think they will be pleased with what they see. They may well wonder if we have made much of a spiritual advance since the time when our ancestors used to occupy themselves mainly with scratching crude drawings on walls.

14

'HERE IS THE ASTROLOGICAL FORECAST . . .'

Now and then, one has a day in which everything seems set to go wrong. I can remember one such occasion not long ago. For no particular reason I woke up feeling un-utterably depressed. Rain was pouring down; the outlook was one of gloom; and I was faced with the prospect of a trip to London, which is very far from being my favourite city. (I vastly prefer the calm of Selsey, where the roar of internal combustion engines is replaced by the mild squawks of seagulls.)

To see what was likely to happen, I consulted the astro-logical columns in several daily papers. No. 1 was quite encouraging. I was going to have important business con-tacts; it was a good day for breaking new ground, and I was going to meet someone who would be a great help to me in my negotiations. Feeling better, I turned to Paper No. 2. This was not so cheerful. Indeed, it seemed that unless I took great care I would have a serious tiff with a loved one. Turning to Paper No. 3, I found that it would be folly to go to London at all, because the omens were highly inauspicious. To enter into any business agreement would mean financial embarrassment at a later stage; moreover, it was on the cards that misunderstandings with associates would result in bitter quarrels and heartbreak. Worst of all, perhaps, I would have to watch out very carefully in case of an accident while travelling.

Actually, nothing much happened. I made a telephone call or two, postponed my appointment in London, answered some letters, continued with a book I was in the middle of writing, and spent the evening at the local cricket nets practising my unorthodox brand of leg-spin in preparation for the morrow's match against West Wittering. (By then, I may add, the rain had stopped, and the Sun shone down from a blue if somewhat watery sky.) But it is true to say that the newspaper astrologers had been of very little help. Anyone who tried to follow them closely would be certain to end up in a mental fog.

Mind you, there are various kinds of astrologers. The "technicians" take themselves very seriously indeed. They hold congresses, issue certificates, cast horoscopes, and present each other with degrees such as D.F.Astrol.S. and F.Astrol.Soc. They look down with scorn upon both the newspaper columnists and upon the professors of astrology who operate from seaside piers. There is, in fact, a major split in the astrological ranks—comparable with that among the flying saucerers, where the U.F.O.-spotters have little time for the Adamskis, the Allinghams or even the globular beings who sit in the Interplanetary Parliament upon Saturn. It is not too much to say that by now there is an element of astrological snobbery. A D.F.Astrol.S. will curl his lip and sneer at the very mention of a crystal ball or a dark lady coming over the water.

Astrology is very old indeed, and it has often been claimed that its very antiquity means that it must be valid. I agree that it is at least as old as Father Christmas, and probably a good deal older; also that it still has a greater following than any other branch of Independent Thought. While the Saucerers, the Flat Earthers, the Hollow Globers and the Velikovskyites can be counted in their tens, their hundreds or their thousands, astrological devotees run into millions. This is particularly so in the Far East. For instance, in March 1971 the Prime Minister of India, Mrs. Gandhi, called an important General Election on a date specifically recommended by her personal astrologer. On this occasion, at least, the astrologer had done his homework excellently; Mrs. Gandhi scored a runaway victory at the polls. I doubt whether a British General Election, for example, would be timed for similar reasons, but there are plenty of British astrologers. You will find them everywhere, from circus tents and seaside jetties to consulting rooms and broadcasting studios.

Up to around the seventeenth century, astrology and astronomy were regarded as equally important, and most astronomers were also astrologers. Since then there has been a definite divergence, and the astrologers have turned more and more to people such as psychologists and psychiatrists, which is perfectly understandable. One enthusiast was a gentlemen named C. G. Jung, who was Swiss by birth, and is regarded as one of the founders of modern psychology. He was, for a time, a follower of another psychologist, one Sigmund Freud—though eventually they disagreed about the rôle of the ego, the id, the subconscious, the semi-conscious, the unconscious or something equally potent, and went their separate ways. I am not sure what Freud thought about astrology, but Jung practised it, and continued to do so until his death at an advanced age in 1961. Another devotee of astrology was Adolf Hitler, and it is said, with a great deal of supporting evidence, that his actions in the late summer of 1939 were influenced by the fact that all astrologers had forecast that there would be no war.

I never met Herr Jung, and I never met Herr Hitler, but I have met many official astrologers (of both varieties), and they always strike me as being among the most single-minded of the Independent Thinkers. Their cult is based entirely upon the apparent positions of the Sun, Moon and planets in the sky, and they use the so-called fixed stars as a kind of backcloth. The stars are not genuinely fixed; they are moving around in space in all sorts of directions at all sorts of speeds, and the star-patterns or constellations have no real significance, because they are purely man-devised. Astrologically, however, this does not matter in the least, and all that has to be done is to work out in just what directions the planets happened to be at the moment of the subject's birth. When this is known, a complicated chart or horoscope can be drawn up, and far-reaching conclusions can be drawn with regard to the subject's character, destiny and even personal appearance.

Take myself, for instance. (And why not, pray? I too am human, despite occasional suggestions to the contrary.) I was born at Pinner, in Middlesex, England, at 10 a.m. G.M.T. on March 4, 1923. Unlike Glendower's, my birth was not accompanied by any celestial portent apart from a violent thunderstorm, which I was told about later even though I do not remember it myself. At that time— March 4—the Sun was in the astrological sign of Pisces, the Fishes, while Venus, Mars, Jupiter, Saturn *et al* were at various spots elsewhere in the Zodiac. Armed with this information, an astrologer can proceed with a horoscope. Bear in mind that it is not only the time of birth which matters, but also the place. Had I made my initial entry in, say, Chipping Sodbury or Ashby-de-la-Zouch instead of in Pinner, my character and career would have been different.

Pisces is not much of a constellation. It consists of some faint stars near the much more important Square of Pegasus, which, however, can be discounted because it is not in the Zodiac (that is to say, the belt around the sky in which the Sun, Moon and bright planets are always to be found). Yet at the time of my birth, the Sun was not in the constellation of Pisces at all. Because of what is called precession, due to a regular alteration in the direction of the Earth's axis, the astrological signs are now out of step with the constellations, so that when the Sun is in the sign of Pisces it is actually in the constellation of Aquarius, the Water-bearer. When it is in the constellation of Scorpio, the Scorpion, it is in the sign of Sagittarius, the Archer; and so on. There is also a very inconvenient constellation called Ophiuchus, which actually butts into the Zodiac for some distance. Astrologers have a positive hatred of Ophiuchus; they pretend he doesn't exist, and they will have no dealings with him.

I hope I make myself clear? Then let us proceed.

The planets are much closer than the stars, which is why they seem to potter around the sky instead of staying in virtually the same relative positions. According to astrologers, each planet has its own special influence, which varies according to its position in the sky and whether it was rising or setting at the time of the subject's birth. This is why the exact time and place of birth is so important. You have only to consider the consequences of Venus being high up instead of low down, or vice versa!

Also, each Zodiacal sign has its own characteristics, and these affect the careers of anyone born under them. For example, a baby born under Scorpio (October 24 to November 22) may well become a psychiatrist, a butcher or a sewage worker; anyone controlled by the sign of Aries (March 22 to April 20) is more probably destined to be a metal worker, an engineer or a trade union leader. And if you are under Libra (September 24 to October 23) possible careers for you include those of a beautician, a diplomat or a high-wire juggler. (Whether these two latter are essentially identical must be a matter for debate.)

141

Astrological interpretations can be continued almost ad infinitum. For instance, each planet is linked with some part of the human body. Neptune controls the nervous system, and in particular the thalamus; Saturn is associated with the skin, teeth and bones; Jupiter with the pituitary gland; Mercury with respiration; and Uranus with the circulation of the blood. As for Mars—well, perhaps we need go no further at present; suffice to say that Mars controls some parts of our bodies which are not only important, but are pre-eminent in our thoughts for much of the time.

All this is quite fascinating, and it would be very useful

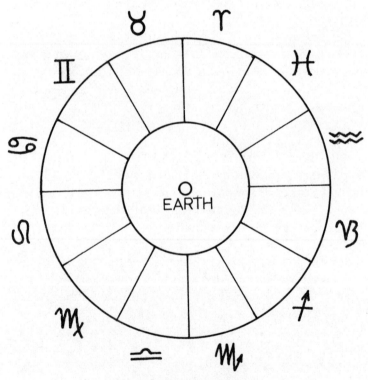

Fig 14 The astrological sky

if a horoscope could predict exactly what course one's life will follow. Unfortunately, matters are not so simple as that, as the astrologers are quick to stress (I refer here to the D.F.Astrol.S.'s, not to the pier-head seers). And in any case, there is one burning question which occurred to me a long time ago. If astrologers are right in saying that astrology works—then *why?*

The names of the planets and the constellations are quite arbitrary. The names we use today were also used by the Greeks, over 2,000 years ago, but few of the constellations bear any resemblance to the objects they are meant to depict. Taurus, the Bull, is fairly typical. I have given a

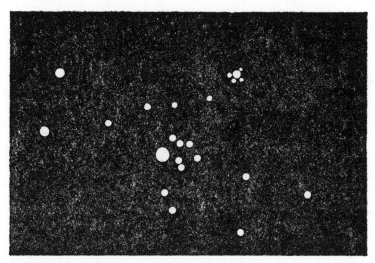

Fig 15 Taurus

map of its chief stars here, and I defy anyone to make a bull out of it. Yet to the astrologers, Taurus is masculine, strong and bull-like—quite different from my own sign of Pisces, which is, I am afraid, watery and anæmic. Pisces is Latin for "fishes", and hence the association with water; but again, it does look rather as though the astrologers first allotted names to the star-groups at random and then calmly assumed that the names had real cosmical significance. The same can be said of the planets, though Mars, the war-planet, is at least red.

This was a question which I put some time ago to a Mrs. Hone, then President of the Astrological Society. She dismissed my objections scornfully, as being totally unworthy of refutation. She also pointed out that no horoscope can be infallible, as the relevant data are almost never complete. In fact, one cannot draw up a chart and then deduce that the subject is left-handed, used to be snooker champion of North Nibley, dislikes asparagus and had a maiden aunt in Hoxton. One must be content with more general and nebulous conclusions.

Having failed to impress Mrs. Hone, I began to check up on astrological predictions of the past. One columnist had forecast the assassination of President Kennedy with absolute accuracy, which was hailed as a major triumph and which quite obscured the fact that in successive previous years he had similarly forecast the untimely and violent deaths of General de Gaulle, President Franco and the Emperor of Abyssinia. (After all, one cannot be right every time.) C. G. Jung, our eminent psychologist, once carried out a long statistical analysis of astrological forecasts, and came to some conclusions which I would no doubt find highly illuminating if only I could make head or tail of them.

I have never quite fathomed how there can be much of a connection between a planet and the background of stars against which it happens to be seen. Throughout much of 1971, the planet Jupiter was seen against the pattern of stars which is known both to astronomers and to astrologers as Scorpio, the Scorpion. The stars in Scorpio are not genuinely associated with each other, and lie at very different distances from the Earth; Jupiter, as a planet in our Solar System, is very much closer than any of them. If we give a scale model, and represent the distances between the Earth and Jupiter by one inch, then Antares, the brightest star in Scorpio, will be roughly 100 miles away. So it is not really logical to claim that during 1971 Jupiter was "in" Scorpio, any more than a bird flying against a background of high clouds can be said to be "in" the clouds.

I put this point of view to two astrologers of very different types. The first one operated from a South Coast pier; I talked to him during a radio interview sometime in 1968. In his opinion astrology depends for its success upon Esoteric Vibrations, fully described in the Ancient Teachings. He also mentioned Atlantis, and said that the only people today who could really answer my questions were the advanced beings who visit us in their flying saucers. (Again, note how all branches of Independent Thought come together in the end.) Unfortunately no Saucer pilots happened to be with us, and I retired baffled.

In a television programme in which I took part, we were joined by Mrs. Julia Parker, D.F.Astrol.S., whom I know very well, and of whose integrity and dedication there is not the slightest doubt. She said, without hesitation, that my question could not be given an answer in scientific terms, and that nobody knew just why astrology worked. To her, a horoscope was a general guide, to be used in much the same spirit as a medical diagnosis. She also said that the future of astrology was certain to be independent of astronomy, and would be much more closely linked with the psychiatrists and psychologists. This certainly makes sense. I wonder whether we shall ever have a new profession—that of astrolopsychiatrist? It does not strike me as being out of the question.

I must end this chapter on a note of disappointment. Only yesterday I read the *Sunday Mirror* astrology column, and found that anyone born under Pisces might well have a week which would be financially profitable. When the post came in on the following morning I opened it with ill-concealed eagerness. Unfortunately there were no cheques waiting for me—only an electricity bill for fifteen pounds and twenty-five new pence. I can only conclude that on this occasion one or more of my guardian planets had blown a fuse.

15
FLOODS, QUAKES AND ICE-BALLS

Have you ever considered building an ark?

I do not, of course, mean anything on the line of Noah's. Today, primitive materials such as wood would certainly be replaced by plastic or something equally sophisticated, and no twentieth-century ark would be expected to carry a full menagerie. But if Dr. Adam Barber of Washington, D.C., is to be believed, all of us will need arks in the foreseeable future. Within the next few decades the Earth will tip drunkenly on to its side; the seas will sweep across the continents, and we will experience a flood beside which the Noahian deluge or the Atlantis inundation would look like nothing more than a burst water-main.

Dr. Barber's followers are, naturally, very ark-minded, and several vessels have actually been built ready for the great day. Dr. Barber himself has written a book on the subject, *The Coming Disaster Worse than the H-Bomb*, and he is not alone in his views. For instance, Mr. Hugh Auchincloss Brown has also been warning about flood dangers, for much the same reasons.

It is all very disturbing, particularly to non-swimmers such as myself. (I live within a few dozen yards of the sea, and frequently go into it, but I would describe myself as a born paddler; this is a pleasant pastime, despite the risk of having one's toes nibbled by shrimp.) So let us consider the situation carefully, prefacing it with a brief description of some earlier forecasts about the approaching end of the world.

Astrologers are particularly prone to such fears. They become highly alarmed when several bright planets cluster together in the same part of the sky, for instance. This is not to say that the planets are really close together; it means only that they happen to be seen in much the same direction, and to an astronomer such a lining-up is considerably less important that a deep depression approaching us from Iceland. To an astrologer, however, it matters a great deal. One famous instance was that of the year 1186, and the astrologers played it up until they caused widespread panic across most of Europe. "In the month of September there will be great tempests, earthquakes, mortality among men, seditions and discords, revolutions in kingdoms, and the destruction of all things," wrote one contemporary astrologer with great solemnity. Nothing happened; and gradually the episode was forgotten.

The next similar scare was that of 1524. Again the cause was a clustering of planets—this time Mars, Jupiter and Saturn. All three happened to be in the constellation of Pisces, the Fishes. Obviously this meant that a flood must be due; with three of the most brilliant planets in the "watery" constellation, what else? So thought Johann Stöffler, the most celebrated astrologer of the day. Warnings went out far and wide, and the resulting panic dwarfed that of 1186, though it is fair to say that England was unaffected. Once again nothing happened, and I mention the incident here only because it led to the building of another ark. It was made at the orders of President Auriol of Toulouse University, who had implicit faith in Herr Stöffler. After the scare was over, the President explained to his colleagues that he had merely wanted the boat in order to go on a fishing holiday.

Passing on, we come to 1919 and the prognostications of an Italian, Professor Porta (I have never been able to find out what he was a professor of—I am sorry, of what he was a professor). The main planets were roughly aligned, and Porta believed that the magnetic currents circling between them would "pierce the Sun like a mighty spear", so that there would be solar storms violent enough to cause floods and earthquakes at home. Finally there was the planet-clustering of 1962, which caused considerable alarm among astrologers in India and Pakistan. As we have seen, however, all was well in the end—thanks to the untiring efforts of the Aetherius Society!

Modern end-of-the-worlders seem to be almost purely religious. Among favoured dates are 1975, 1984 and 1999. Recently I saw a procession to this effect; it was in London, winding its way along Oxford Street in a perfectly peaceful manner. As I watched, the leader, a bearded patriach bearing a banner "Prepare to Meet Thy Doom!" was knocked down by a passing bicycle. He was unhurt, and was able to rise to his feet in time to unleash a stream of the most lurid invective at the departing cyclist. According to my rough estimate he kept on for at least five minutes, saying different things all the time. But to delve into the religious aspect would be beyond my scope; so let us come back to the true Independent Thinkers, and consider Dr. Barber in more detail.

He is by profession a gyroscope engineer, and his flood theory is based upon the behaviour of a gyroscope. As everyone knows, a toy gyroscope will behave in a most diverting manner just before it topples; it waves to and

Toppling gyroscope

Fig 16 A toppling gyroscope

fro, and its "pole" will sweep out a circle. The Earth does much the same thing, and is wobbling as it spins. Each wobble takes 25,000 years; the phenomenon is called precession, and it is this which has led to the Zodiacal signs getting out of step with their constellations.

Dr. Barber began his studies while trying to make a perpetual motion machine. (He has, indeed, made no fewer than seventy-five of these, though none of them has actually worked.) He next tried to build a gyroscope which would give a small-scale reproduction of the Earth's move- ment round the Sun. And then came the great discovery which made him rock on his heels and gulp for breath. The world has a "small orbit" which astronomers have overlooked!

According to Dr. Barber, the length of the Earth's annual path round the Sun is 572,615,372,493,297 miles. His gyroscopic "small orbit", spread out over the large one, is 9,865,621,106,441,698,602 miles (I hope nobody is going to ask me just how he arrived at this latter figure.) He goes on to say that when the Earth's axis makes a right angle with both the large and the small orbits at the same time, the axis will shift suddenly by 135 degrees, so that the North Pole will shift by 90 degrees. The danger-times are June 21 and December 21 each year, when the Earth reaches the extremes of its orbit.

I hope this is clear? Just in case it is not, let me quote Dr. Barber's own words:

"The shift is basically caused by gyroscopic pressures at right angles to the two orbits of the Earth. This occurs because the Earth is a huge gyroscope, and when it reaches the dead centre of its orbits it can no longer press forward. On account of its momentum in its orbit, it creates a gyroscopic pressure at right angles, thereby causing the sudden shift.... The shift will be completed in about $1\frac{1}{2}$ hours."

According to Dr. Barber, this shift happens every 9,000 years or so. The last occasion was (naturally) at the time of Noah. Another shift is due; and we must all prepare for it.

Dr. Barber's book makes fascinating reading, and is essentially scientific. He writes, however, that on one occasion during his researches he had the personal assistance of God. He was in need of a coil spring (such as one from an old gramophone) for one of his pieces of apparatus. Pondering upon the difficulty of finding one, he set out for the junk-yard near his home. Half-way there he tossed a cigarette-end into the gutter, and—*mirabile dictu!*—it landed right on a gramophone spring of the kind he wanted. What could be clearer proof of divine guidance?

The first edition of *Disaster Worse than the H-Bomb* appeared in 1954, and caused considerable interest. Copies were sent to all leading institutions; to scientists of all disciplines; to Governments; and to personalities such as Sir Winston Churchill, General de Gaulle, the President of the United States and the Emperor of Japan. In general, it is fair to say that astronomers were not impressed. Letters from the Royal Greenwich Observatory, the U.S. Naval Observatory and the Mount Wilson and Palomar Observatories were reproduced in Dr. Barber's second edition (1957), and were somewhat scathing. However, periodicals such as *Fate, Chimes, Search* and *Psychic Observer* were much kinder, and Dr. Barber proceeded to the establishment of the Barber Scientific Foundation.

There were various reasons why the Foundation had become necessary. First, there was the question of a warning system. Dr. Barber recommended hanging a large ball from the ceiling in one of the living-rooms of each house. If the Earth began to tilt, the ball would sway, so closing an electric circuit and making a bell ring. This would allow everyone within range to climb aboard a boat which, presumably, had been moored nearby.

Why a boat? Surely the answer is obvious. When the axis tilts, the seas will be "left behind", so to speak, and the oceans will roar over the lands. Fortunately, Dr. Barber assures us that the effect will be gradual. The first sensation will be a mild surge, continuing for several minutes. Then, slowly, the tilt will increase; the waters will swirl over cities, forests and deserts, and all will be chaos. Only after an hour and a half will things calm down again. The tilt will be over, with a total change in the Earth's geography, but with the comforting knowledge that nothing much else will happen for another 9,000 years.

If enough boats are to hand, casualties can be kept to a minimum; this is why Dr. Barber has sent copies of his book to statesmen as well as to scientists. He suggests mooring arks on every street-corner, because the shift may now be expected any year. Of course, he has himself taken full precautions, and has built an ark tough enough to cope with any situation.

Another bright idea involves balloons. What one does is to construct a large gas balloon and moor it to one's front gate. At the start of the axial tilt, simply shin up a rope ladder into the safety of the wicker basket attached to the balloon. Keeping the balloon inflated during the waiting period may be a problem; but, after all, the whole theory is associated with so much hot air that the difficulty can doubtless be overcome.

And yet is there anything that we can *do* about this? It seems a pity to wait supinely until we are swamped; and Dr. Barber has come up with some practical suggestions. He proposes to fit large reaction jets to the tops of mountains on opposite sides of the world (see diagram). When

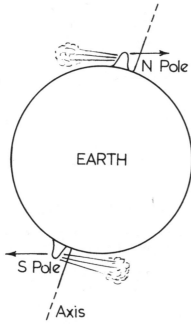

Fig 17 Dr. Barber's jet plan

the tilt begins, one simply switches on the jets; the power produced jerks the Earth back into its original position, and nothing will be felt apart from a slight shudder as though the world had been prodded with a gigantic fork.

During the decade and a half since Dr. Barber's revelations first hit the headlines, there has been no axial slide; but the Foundation is still in existence, ready to help and succour those who need it. Dr. Barber is himself the President, Director and Managing Trustee. I do not think there are any other members, but one is clearly enough.

A variant upon the main theme has been provided by Mr. Hugh Auchincloss Brown, an electrical engineer of Douglaston, N.Y. He too anticipates a flood, but his reasons are not quite the same as Dr. Barber's. In his view, the danger comes from the growing weight of ice at the Earth's South Pole. The ice-cap is growing; when it reaches a critical value the world will become top-heavy, slipping over sideways in a more or less Barberian manner. His solution is to cut channels in the ring of coastal blocks forming the basin which holds the great ice reservoir. When this has been done, the force of gravity will cause the central glacial ice to drain peacefully off into the oceans.

Operations such as this—or even, for that matter, Dr. Barber's arks and jets—need the support not only of Governments, but also of international organizations. One looks automatically at the United Nations. I am told that once, some years ago, there was a resolution about something or other upon which all the U.N. delegates agreed, and passed unanimously. I find it hard to believe that the United Nations has ever agreed about anything, and certainly there is no delicate situation which cannot be made worse by an official U.N. resolution; but even if this were not so, I rather doubt whether such projects would find approval there. Independent Thinkers have to remain independent. Occasionally their ideas are discussed at political level; we have "Symmes' hole", Mr. Gladstone's abortive efforts to raise funds for an Atlantis expedition, and the extraordinary Project Habbakuk, which, to me, remains the weirdest episode of the whole war. But such cases are exceptional.

Meantime, another Independent Thinker has come up with a theory which introduces gyroscopes. Mr. P. Norcott, of Broadstairs, England, has been studying the cause and possible prevention of earthquakes; he has written a booklet, and has also explained his ideas on the radio. Unlike most other astronomers, he believes that the Earth was formed by material blown out from a sunspot. He agrees that it is an oblate spheroid (that is to say, somewhat flattened; the equatorial diameter is 27 miles greater than the polar) and he also agrees that its rate of spin is slowing down, so that the "day" is becoming longer. Though the rate of slowing-down is slight, and amounts to less than a thousandth of a second per century, Mr. Norcott feels that it is significant. Because the centrifugal force around the equator is decreasing, pressure is building up below the Earth's crust; eventually something has to give way, and—bing!—we have an earthquake.

Earthquakes are highly destructive, and anyone who can find a way to prevent them will be performing a great service to mankind. Mr. Norcott believes that he has the answer. What we must do is to stop the steady slowing-down of the Earth's rotation, and this can be achieved by using gyroscopic force. Build huge flywheels, erect them at the North and South Poles, and start them whirling. They will keep "kicking the Earth round", so to speak, and will keep the rate of axial spin constant.

There is another method, however. The slowing-down of the Earth's rotation is due to the tidal influence of the Moon; the principles have been known for a long time, and there is nothing mysterious about them. By building huge dams to stop the ocean waters sloshing about from one sea-basin into another, tidal friction could be reduced admittedly at the cost of some inconvenience; yachtsmen would object to having their routes blocked, and the Navy would also have to be consulted. Mr. Norcott's third possible solution is to split the Moon in half, towing one hemisphere of it round to the other side of the Earth so that in future the gravitational forces will cancel each other out. Unfortunately this might be rather dangerous,

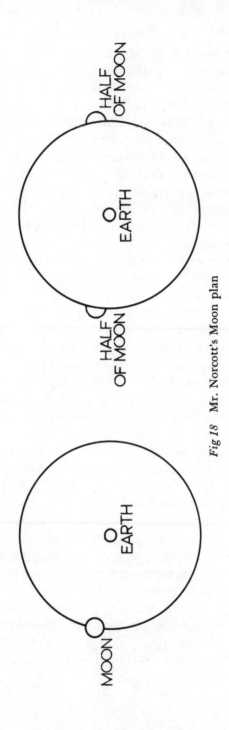

Fig 18 Mr. Norcott's Moon plan

as Mr Norcott is quick to point out. Nuclear power would have to be used to disrupt the Moon, since old-fashioned methods would be no good at all. One hates to think how long it would take a man with a hacksaw to cut right through the lunar globe.

Dr. Barber and Mr. Hugh Auchincloss Brown are concerned with a possible flood disaster, and each has proposed methods of dealing with the crisis when it comes. Mr. Norcott is concerned with disasters, and he too has made suggestions which are intriguing even if rather hard to put into practice. All three are very far removed from the gloomy End-of-the-Worlders who have so often appeared in the past, and will no doubt continue to do so in the future. And to conclude the present chapter, it seems fitting to give a brief account of the equally constructive ideas of Mr. A. P. Pedrick, who lives in my own village of Selsey, and who is a qualified engineer with many original proposals.

Mr. Pedrick has been considering the "waste areas" of the world—places such as the Sahara Desert, Death Valley, and the interior of Australia. Nothing useful can grow in these places, simply because there is not enough water. After the end of the last war, a British Labour Government which contained many Independent Thinkers planted a vast number of ground-nuts in an arid region, and expressed astonishment when all the nuts died (the planted ones, I mean; not those in the Labour Government). But if the desert regions are ever to be made fertile, they must be provided with water.

On the other hand, there are areas in which water is superabundant. Salt water is useless, of course, which rules out the oceans; but there is unsalted water at the polar regions, in the form of snow. Clearly what has to be done is to transport the snow from the poles to the deserts. One can hardly carry it in lorries, and a quick, economical method must be worked out.

Mr. Pedrick proposes to build pipe-lines, and to make use of the Earth's axial spin. First, construct pipes leading from high latitudes to low—in particular, from Antarctica to Central Australia. In the polar stations, equipment is provided for compressing snow into ice-balls. These balls are fed into the pipe-line, preferably at a substantial height above sea-level so that they acquire a velocity through the pipe under the action of gravity. It may also be possible to create a pressure difference across the faces of the balls, so that they are accelerated in the manner of a piston. The movement of the balls through the pipe is then maintained, or further accelerated, by the rotation of the Earth around the polar axis. Finally the balls pop out at the other end of the pipe; they are rolled into a suitable depression, where they melt in the sunlight and release water. All that remains is to pump the water across to the regions where it is needed most.

Mr. Pedrick is a mathematician as well as an engineer, and he has gone to immense trouble to make the necessary calculations. In 1966 he even took out an elaborate patent for the scheme. He finds, for instance, that the entire "Dead Heart" of Australia (about a million square miles) could be irrigated to the extent of 30 inches of water by about one hundred 10-foot bore pipe-lines, with ice balls flowing through at 500 m.p.h. and spaced five yards between the centres of the balls.

He is under no delusions about the engineering problems to be faced. The material of the pipes alone will be hard to choose; after all, much of the track will have to be laid under the sea. Also, in order to reduce drag and friction, the pipes must be pumped free of air as efficiently as possible. In very blustery areas, such as those of the Roaring Forties, floating maintenance platforms are to be provided, powered largely by special forms of windmills. A break in the pipe would obviously be disastrous; it would mean that all the balls would be dumped in the sea.

Unlike Mr. Pedrick, I am not an engineer, and therefore I am in no position to comment upon his ideas. But nobody is likely to deny that they are original; and they represent a very real effort to tackle a problem which is going to become increasingly pressing in the years to come. World population is growing fast; we need all the room we can find—and if we can irrigate our deserts by means of Mr. Pedrick's ice-balls, so much the better.

Meantime, we must await events. I cannot tell whether anyone will build Mr. Norcott's flywheels, or try to split the Moon in twain, and I have not gone to the lengths of building an ark or setting up an alarm system of the type recommended by Dr. Barber. To be quite candid, I do not believe that the Earth's axis will slip in the way that Dr. Barber and Mr. Brown expect. If it does, I shall have to depend upon the rowing-boat in my garden—regardless of the fact that it has no oars, lacks one of its rollocks, and has a hole in the bottom. But at least I cannot pretend that I haven't been warned.

16

WATCH ON THE RHINE

Science fiction is great fun. I have read a great deal of it; I have even written some. One of the favourite themes is that of telepathy, and there are few respectable Martians or Alpha Centaurians who cannot tune in to one's inner thoughts and read them just as easily as you can read the words on this page. It must be a gift which leads to embarrassing moments.

Nobody I know can practise telepathy with any reliability or consistency at all, but all the same there is a great deal of evidence that there is "something in it". Could it ever be developed? During my researches for a television programme, I met a man who has been trying to photograph people's thoughts. His name was George de la Warr; I spent a memorable afternoon with him, and it was with deep, genuine regret that I learned of his sudden death only a few months later.

Of one thing I am certain: nobody could have been more honest or more dedicated than de la Warr. I will come to his theories in a moment; but because they are somewhat intangible, and deal with subjects which are hard to grasp, I must pause first to discuss some other Independent Thinkers who have been experimenting with what, for want of a better term, I must call thought-transmission. And surely one must begin with Dr. Joseph Banks Rhine, of New York, whom I have not personally met, but who has been studying extrasensory perception: E.S.P., or, if you like, telepathy.

Dr. Rhine depends mainly upon cards and dice. By "cards" I do not mean ordinary playing-cards; Dr. Rhine's bear designs such as squares, circles and crosses. After carrying out literally millions of tests, he has found that designs can be transmitted by some mental process from one man (or woman) to another. Also, by thinking hard about one particular number-face on a dice, and then throwing the dice, it is possible to make this number-face turn up more often than it would otherwise do.

The usual procedure is to start with, say, two hundred volunteers. Generally, about half these will have "runs" of above random average, and the other half will have lower scores. Of course, this means that the high-scorers are the more sensitive; so the others are cast aside, and the tests are repeated. Again half the volunteers are eliminated. Eventually, after repeating the process a sufficient number of times, we are left with a few people who have had consistently high scores, and who presumably possess a certain amount of telepathic power.

I have always had an uneasy feeling at the back of my mind that there is a flaw in this reasoning somewhere, but I have never carried out any tests, and so I would not dare to criticize Dr. Rhine's. He is himself quite confident that the results are valid. And so, having demonstrated the existence of thought-waves, let us turn to practical matters.

If thought-waves exist, we must also open the door to radiations and vibrations of many other kinds. Indeed, it may even be that everything, animate and inanimate, sends out radiations of a sort which we cannot detect by orthodox methods. Fifty years ago an American doctor, Albert Abrams, set out to apply this theory to medical diagnosis. Basically, what he did was to take a drop of blood from a sick person, put the drop on to a filter-paper, join up some electrical circuits, fix a wire to the forehead of a healthy person, and then tap the volunteer's stomach. By listening to the sounds produced, Dr. Abrams claimed to be able to diagnose the nature of the disease affecting the patient—who might well be miles away. Of course, it was important to remember that the healthy volunteer always had to stand facing *west*.

Dr. Abrams collected many followers, and, incidentally, made a great deal of money. Even today there are Abrams-type practitioners in the United States. Having diagnosed the cause of the trouble, they proceed to broadcast "healing rays" from special vibratory machines. Not having seen these devices, I cannot comment upon them, but it is worth noting that a Dr. Drown, one of the leaders of this school of thought, has written a book called *Wisdom from Atlantis*.

Then there was Wilhelm Reich, yet another psychologist associated with Sigmund Freud and C. G. Jung. Reich is

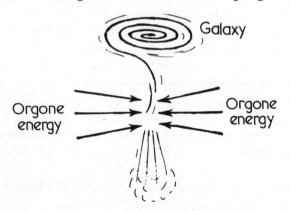

Fig 19   Galaxy formation according to Wilhelm Reich

162

today regarded as one of the most important founders of modern psychiatry, and this I can well believe. He believed in what he called orgone energy, which permeates all nature; it is blue, and is responsible for phenomena such as lightning, the colour of the sea, heat-waves rising from tarred roads, and the luminosity of the tails of glow-worms. It is also responsible for human sexual energy. He built an orgone energy accumulator, made up of a box inside which a patient sat—soaking up orgone energy in much the way that blotting-paper soaks up ink.

Reich even entered the realm of astronomy. In his book *Cosmic Superimposition*, published in 1951, he explained that galaxies, or star-systems, are formed when two amorphous streams of orgone energy rush together, producing matter which eventually condenses into stars, planets, you and me.

I am sorry that I have not been able to lay hands on an orgone energy box to show on television. We did, however, record a long interview with George de la Warr, at the de la Warr Laboratories in Oxford. He died before the programme was scheduled to be transmitted, and obviously we did not include the interview, good though it was. The following notes are drawn chiefly from a transcription of what was actually said at the time.

Biomagnetics, Mr. de la Warr told me, is a new science, and may be defined as being a study of the finer forces of nature. It deals largely with the pre-physical state of matter: that is to say, matter before it actually exists. Much depends upon the fundamental radiations sent out from the object, and upon exact alignment with the critical position in the general magnetic field. When everything is suitably arranged, the object will be *en rapport*, and the fundamental energy will behave in some respects like a light wave; it can be detected by special equipment, and it can even be photographed.

The detecting device involves a framework and a rubber

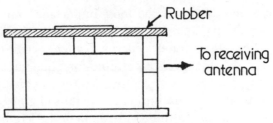

Fig 20 The de la Warr camera

strip. The operator rubs the strip, and when it becomes "sticky" one knows that the object is *en rapport*. The special camera uses an ordinary film, and is able to detect the pre-physical form of the object. Consider, for example, a horse-chestnut. When placed in the camera and rotated to its critical position, its magnetic field will be stimulated with a pattern of energy—and it is possible to photograph the flowers which do not yet exist, but will do so in the fulness of time (unless, of course, someone eats the chestnut first). It is all a question of aligning the object's molecules with the pre-physical energy in the growth position.

The camera itself was a most interesting arrangement, involving magnetic disks, rotating magnets, and a spiral collector. Obviously it could not be expected to function unless everything were precisely *en rapport*, and this meant that the operator had to be specially trained in order to get the best out of it. I tried it myself, but there was no time to check my results, and I doubt whether I would be suitably skilled. (I have a very good camera of the conventional type; it takes clear pictures, but is inclined to cut people's heads off.)

Having established the principle of *en rapport*, we can put it to use. Consider a field containing crops which are not doing at all well, possibly because of poor soil or the invasions of beetles. What can be done is to take photographs of the crops, and then bring the photographs back to be treated in the laboratory. Mr. de la Warr has found

that by treating the photographs, much benefit will accrue
to the crops themselves, no matter how far away they may
be.

This system, if developed, will be of the utmost value to
farmers, but one has to be careful about it, as Mr. de la
Warr found in 1955. As related in his book *New Worlds
Beyond the Atom*, he treated samples of soil from five acres
of a tobacco farm in Rhodesia. In his anxiety to get good
results, and being unable to check the immediate effects—
simply because Rhodesia is rather a long way from Oxford
—de la Warr over-treated the soil samples. Consequently,
the tobacco plants in the Rhodesian farm flowered much
too early in the season, which made things rather awkward.
However, these errors are bound to happen in the early
stages of pioneer work, and no lasting harm was done.

Mr. de la Warr also paid attention to the medical aspects
of biomagnetics, rather as Dr. Abrams had done years
earlier, but in a much more sophisticated manner. In some
cases he found it possible to treat, and cure, a disease which
had not yet appeared, as shown in one interesting case
which had occurred quite recently. He was given a hair
from a man in a London hospital, fifty miles away; by
examining the hair in his special camera, he detected signs
of tuberculosis in one lung. X-ray pictures taken in the
hospital itself showed nothing, and so evidently the disease
had not actually started. By transmitting healing from
Oxford, the situation was dealt with, and in fact the
patient never did show any sign of tuberculosis. If this
method can be developed, there will be many ramifications.
For instance, it will be very reassuring to know that biomagnetics can always cure an attack of influenza that you
haven't got.

Another application of biomagnetics is in mining and
prospecting. Take an aerial photograph of a wide area of
the earth; treat the photograph with the special equipment,
and it will be found that, for instance, certain regions in

the photograph are *en rapport* with water. This shows that underground water will be found on the actual site, in the positions indicated. More impressive still is the idea of tracking down actual people from studying a photograph in conjunction with that of a fingerprint or something equally personal.

When I asked Mr. de la Warr exactly why his experiments worked, he said frankly that he did not know. He was venturing into uncharted territory, and up to that time he had done no more than reach the fringes of it. This is quite logical. After all, remember that when Marconi managed to send the first radio signal across the Atlantic he had not the faintest idea of how he had managed it; all the scientists had told him that nothing of the sort would be possible. All he knew was that the signal had come through. Whys and wherefores could wait until later.

Finally, we came down to the vitally important, albeit rather dangerous, possibility of photographing people's thoughts. Mr. de la Warr told me that in 1950 he had managed to obtain an admittedly rather blurred picture of his wedding-day, which had been as long ago as 1929; he built a special apparatus with a time vector, concentrated hard, and produced something which was identifiable. As he said, "A past event has been photographed. This means that Time itself is a vector of the magnitude spectrum."

Naturally I was anxious to try for myself, and we did expose one plate, but the result was inconclusive. I tried to produce a thought-picture of a Tibetan yak climbing a mountain. Unfortunately the result was badly fogged; and as by this time I was myself in very much of a mental fog, we decided to call it a day.

I do not know what will happen now to biomagnetics. Virtually all the pioneering work was carried out by Mr. de la Warr and his team, and it may well be that there will be something of a lull before anyone else picks up the torch. My own feelings are somewhat mixed. Medical and mining diagnoses would be excellent, but I am a trifle uneasy about thought-photography. On the other hand, there would be one major benefit inasmuch as the whole system of politics would have to be cast aside and re-modelled.

Actually, thought-photography did not originate in Oxford. It was pioneered as long ago as 1910 by a Professor T. Fukurai, of Japan. Placing photographic films or plates in a suitably sealed casket, he thought hard, and impressed Japanese characters on the plates clearly enough to be shown recognizably. This would be a difficult process even with Roman lettering; anyone who can do it with hon. Japanese characters earns my undying respect! Alas, I know of nobody who has followed up Professor Fukurai's work, and neither have I been able to find out what happened to him.

When we pass on to telepathy, we are more or less leaving the realm of Independent Thought, but there is one telepath who certainly qualifies for inclusion here, and who did indeed provide the grand finale to our television programme. He is Mr. Bernard Byron, of Romford, England. We went to see him for a very special reason. Mr. Byron is one of the few Earthmen privileged to be able to speak, and write, languages of other planets. He is fluent in Venusian, Plutonian and Krügerian—the latter being the tongue employed by the people who live on a planet moving round the far-away red dwarf star Krüger 60.

In our previous forays we have always found extra-terrestrials who can communicate with us. True, Allingham's Martian did not speak English, but semaphore provided a good substitute; George Adamski's original visitors were coy, but when they got to know him better they showed themselves to be completely fluent. Aetherius, as we have noted, can speak all Earth languages (except French and Norwegian, of course). Most Saucer pilots and crews are similarly gifted. But it would be too much to expect other beings to use English when chatting among themselves; and this is where Mr. Byron comes into the picture.

I have corresponded with him for some time, and have seen examples of the writing used in these remote worlds. He tells me that the languages have been transmitted to him by rays, which means, in effect, by telepathy. He has also seen the people responsible. Venusians have lovely blue eyes and blond hair; Plutonians are very small, with long hands, very long thumbs, and egg-shaped heads coming upward to a point, while Krügerians have one lung at the top of the chest and the other right down in the body. Because these lungs function independently, Krügerian words tend to be strung together, and the result sounds like a stream of running water. Mr. Byron cannot speak Martian, but he does know that the Martians have their eyes on the sides of their heads.

Mr. Byron was kind enough to speak in all three cosmical languages. The result was absolutely tremendous. I very much regret that the words cannot be written down phonetically; but perhaps one day the Venusians *et al* will mix more freely with us, so that their language will be taught in our schools as widely as French is today. However, it cannot be said that interplanetary languages are easy, and to me they recall sea-lions gargling. If ever we have a cosmical Common Market, I personally will put in a strong plea that for our official tongue we should retain good, old-fashioned English. If need be I am prepared to accept French, German, Swedish, Urdu, Hindu, Voodoo or Sanskrit; but Plutonian—no. Neither do I propose to make any attempt to learn it myself. I know my limitations.

And this, I think, brings me for the moment to the end of my account of Independent Thought. I wonder what your impressions are? Remember, yet again, that when men such as Copernicus, Dalton, Tsiolkovskii and Baird first put forward their views they were regarded not only as unorthodox but also as highly eccentric. To anyone living in the time of Henry VIII, the idea of a moving Earth seemed much wilder than Mr. Francis' present-day theory of a cold Sun. Baird said that he could send pictures through the air; and people laughed at him. Rhine and de la Warr say that they can send thought-pictures through the air; is it wise to laugh at them too?

I began this book by admitting that I am a conventionalist, reluctant to accept revolutionary ideas. This is ingrained in me, and in many other people too. I would say that a good 99 per cent. of mankind is prepared to accept what the textbooks say. We have been discussing the theories of the remaining one per cent.

Usually, as we must concede, the Independent Thinker is wrong; but by his very nature, he is also apt to be devastatingly right when others ridicule him. And whether right or wrong, he has an invaluable contribution to make to this modern world of ours. His refreshing, thoroughly healthy revolt against orthodoxy makes other people concentrate hard; and from this, nothing but good can emerge.

And so, in conclusion, let me pay tribute to the Independent Thinkers. To see the last of the Flat Earthers, the Cold Sunners, the Anti-Evolutionists, the astrologers, the Velikovsyites, the Flying Saucerers and their kind would be tragic, but fortunately I am sure that it will not happen. In the centuries to come, mankind will still be producing its men of Independent Thought, just as it has always done in the past. We welcome them—and even if we do not agree with them, we salute them and their courage.

169

INDEX

171